LAKE CLASSICS

Great American Short Stories I

Sarah Orne JEWETT

Stories retold by C.D. Buchanan
Illustrated by James Balkovek

LAKE EDUCATION
Belmont, California

LAKE CLASSICS

Great American Short Stories I

Washington Irving, Nathaniel Hawthorne, Mark Twain, Bret Harte, Edgar Allan Poe, Kate Chopin, Willa Cather, Sarah Orne Jewett, Sherwood Anderson, Charles W. Chesnutt

Great American Short Stories II

Herman Melville, Stephen Crane, Ambrose Bierce, Jack London, Edith Wharton, Charlotte Perkins Gilman, Frank R. Stockton, Hamlin Garland, O. Henry, Richard Harding Davis

Great British and Irish Short Stories

Arthur Conan Doyle, Saki (H. H. Munro), Rudyard Kipling, Katherine Mansfield, Thomas Hardy, E. M. Forster, Robert Louis Stevenson, H. G. Wells, John Galsworthy, James Joyce

Great Short Stories from Around the World

Guy de Maupassant, Anton Chekhov, Leo Tolstoy, Selma Lagerlöf, Alphonse Daudet, Mori Ogwai, Leopoldo Alas, Rabindranath Tagore, Fyodor Dostoevsky, Honoré de Balzac

Cover and Text Designer: Diann Abbott

Library of Congress Catalog Number: 94-075019
ISBN 1-56103-009-0
Printed in the United States of America
1 9 8 7 6 5 4 3 2 1

THE LANGUAGE OF LIFE

How Cells Communicate in Health and Disease

Debra Niehoff

Joseph Henry Press
Washington, D.C.

Joseph Henry Press • 500 Fifth Street, NW • Washington, DC 20001

The Joseph Henry Press, an imprint of the National Academies Press, was created with the goal of making books on science, technology, and health more widely available to professionals and the public. Joseph Henry was one of the founders of the National Academy of Sciences and a leader in early American science.

Library of Congress Cataloging-in-Publication Data

Niehoff, Debra.
The language of life : how cells communicate in health and disease / Debra Niehoff.
p. ; cm.
Includes bibliographical references and index.
ISBN 0-309-08989-1 (cloth)
1. Cell interaction—Popular works.
[DNLM: 1. Intercellular Signaling Peptides and Proteins—physiology. 2. Paracrine Communication—physiology. QH 604.2 N666L 2005] I. Title.
QH604.2.N54 2005
611′.0181—dc22

2005002386

Cover design by Michele de la Menardiere; Nerve Cell Culture image © SPL/Photo Researchers, Inc.

Text illustrations by Michael Linkinhoker, © Link Studio LLC.

Printed in the United States of America.

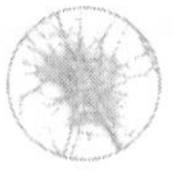

CONTENTS

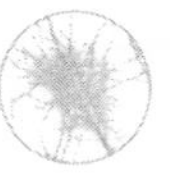

INTRODUCTION

One of earth's most primitive civilizations isn't hidden in the depths of a tropical rain forest or sequestered on a remote island. It has no artifacts to analyze, no traditions to document, no exotic rituals to preserve. It lives in pond scum, and its name is *Volvox.*

Under the microscope, magnified fortyfold, *Volvox cartieri* (its full name) resembles an olive-colored golf ball, stamped with a handful of dark green spots. At higher magnifications, its dappled "skin" resolves into a mosaic of cells, embedded in a sphere of translucent jelly, the spots into clusters of daughter cells clinging like soap bubbles to the inner wall of the sphere. Trees and grasses are anchored by their roots, but *V. cartieri* can swim and spin, propelled through the water by whiplike projections, or flagella, that sprout from each of its cells.

The family Volvocacae—the tribe of green algae that includes *V. cartieri*—are a sociable lot. In contrast to the independent lifestyles of unicellular organisms that biologists call "protists," many volvocacaens prefer to live in groups, ranging in size from tiny *Gonium* with a dozen-odd cells to *Volvox* species containing as many as 50,000—a veritable city of an organism. But a crowd is not necessarily a community. The cells that comprise *Gonium*, for example, are no more than guests in the same motel. Anonymous, transient, and opportunistic, each can decide at any time to quit the group and start its own colony. *V. cartieri*, on the other hand, has adopted a social contract similar to that honored by plants and animals. Its cells have relinquished their autonomy and pledged lifelong commitment to each other. No longer capable of living alone, they die if separated from one another.

V. cartieri is indivisible because it has opted for a division of labor. One type of cell, designated "somatic," is responsible for infrastructure and transportation. These cells keep a roof over everyone's head, as well as build and operate the flagella that power the colony around the pond. Others, known as "gonidia," have no flagella. Confined to the southern hemisphere of the cellular globe, they specialize in reproduction; the daughter colonies are their handiwork. But their most celebrated reproductive skill is sex. Come midsummer, they undergo a primitive sort of puberty in which some—indistinguishable from their colleagues on the outside, genetically male on the inside—differentiate into sperm cells, while the others, secretly female, produce eggs. At maturity these germ cells abandon the soon-to-die somatic cells and mate, giving rise to a resilient zygote able to weather summer's heat and winter's drought by taking refuge in the silt at the bottom of the pond, where it lies dormant until next year's spring rains.

Gonium is a coalition. *Volvox*, on the other hand, is a collaboration, tens of thousands of cells forged into a "whole functioning interdependently"—the *Oxford Dictionary* definition of a society. Such teamwork is possible only because the members of *Volvox* have evolved one additional, priceless faculty: language.

By pooling resources and dividing the work, a human society can accomplish things that would be impossible for a single individual:

construct a city, feed the masses, maintain an army, build an empire. Cooperation and specialization also allow a cellular society—even one as simple as *Volvox*—to exploit natural resources and cope with emergencies in ways that are simply beyond the reach of a single cell. For example, because it has mastered the art of synchronized swimming, *Volvox* can glide quickly and efficiently to the choicest spot in the pond. And should the pond threaten to dry up in the summer heat, *Volvox*'s discovery of sex and the marriage of specialized germ cells provide for the survival of the next generation (and for the family genes the young inherit from their parents), even if every last somatic cell dies. In more advanced multicellular organisms, like humans, an even greater degree of specialization and organization coordinates the action of millions of cells to form the cellular equivalent of states—highly stratified, complex societies that have not only adapted to the challenges of environments as diverse as the rain forest and the desert, the open sea, and a backyard pond but have also helped shape those environments.

Cooperation requires conversation. As biologist John Tyler Bonner puts it, all biological societies have in common the need for "some coordination, some integration or communication between its members." Humans speak to one another. Sounds, scents, and postures connect members of animal societies. *Volvox* cannot talk, hiss, growl, trill, or show its teeth, but even the members of this simple cellular society have found a way to communicate. Strands of cytoplasm connecting its constituent cells transmit chemical substances that coordinate the beating of the flagella, while the somatic cells, unable to participate in sex, say the word that initiates it. Sensorium and sibyl as well as transport coordinator, they read the signs—rising temperatures, less room to move, a subtle brackishness of the water—and predict the future. Before dehydration overtakes them, they warn their fecund sisters, producing and secreting a protein that diffuses into the surrounding water and triggers the sexual differentiation of the gonidia, the reproductive cells.

The language of all cellular societies is similarly based, not on sounds or gestures but on chemistry. Using molecules where we would use words, constructing sentences from chains of proteins, the cells that make up the bodies of multicellular organisms inform, wheedle, command, exhort, reassure, nurture, criticize, and instruct each other. Their conversations direct every physiological function, report every newsworthy event, record every memory, heal every wound. Like our spoken and written language, which can be used to describe both the past and the present, or fact as well as fantasy, this chemical language is versatile, as useful for orchestrating the development of tissues and organs as issuing a request for more food. Like our language, it is flexible, permitting molecular "words" to be combined in more than one way, as well as accommodating, open to the addition of new words. Our language—aided by the postal service, the telephone, the fax machine, and the Internet—enables correspondents separated by long distances to communicate with one another. Similarly, diffusible chemical signals, relayed from sender to recipient via the bloodstream or directed over long distances by cellular structures that are the equivalent of telecommunications networks, permit even distant cells to share information, so that a biological society, regardless of its size or complexity, can coordinate the activities and regulate the social behavior of its many members. Just as many social scientists consider language to be one of the seminal accomplishments of human society, biologists rank chemical communication as one of the most important advances in cellular societies, a skill essential to the evolution of the multicellular lifestyle.

The molecular biologists who worked for over a decade to sequence the human genome have sometimes referred to that sequence as "the book of life." But even they acknowledge that our DNA alone actually tells us little about the way our bodies operate. To our cells, the book of life is no more than a reference manual; it's the proteins encoded in the genes, not the genes themselves, that actually build and maintain the molecular machinery essential to life, including the network of signaling pathways that allow our 60 trillion cells to function as a single organism. Genes conserve, but only living cells can converse; from the words written in their genes over billions of years of evolution, they construct a language—the language of life.

Cells had barely been discovered when scientists peering into primitive microscopes first began to examine the architecture and activities of their civilization. In 1682—only two decades after the first published description of plant cells—Anton van Leeuwenhoek (the Dutch lens grinder and amateur microscopist famous for his descriptions of one-celled "animalcules") wrote a letter to the Royal Society documenting his observations of a structure known today as the cell nucleus. In the 1820s a French physician, Francois-Vincent Raspail, incinerated cells on a platinum spoon and analyzed the residue to discern their chemical makeup. By the end of the nineteenth century, cell biologists had discovered dyes and staining techniques that revealed new aspects of the cell's internal structure; within a few more decades, they'd learned how to deconstruct cells and separate their constituent components by spinning the slurry in a centrifuge. By the end of the twentieth century, the electron microscope, advances in protein chemistry, radioactive tracers and fluorescent markers, and the techniques of molecular biology had described the structure, physiology, and genetic makeup of cells with exquisite detail.

The excavations carried out by anatomists, biochemists, and molecular biologists yielded precious artifacts: the tools used by cells to make a living, the features of their personalities, the machinery that furnishes their interiors. And they also revealed molecular words and protein phrases, fragments of the most ancient language on earth.

These discoveries, as important to biology as the discovery of the Rosetta stone was to history and linguistics, have added to the story of life the direct testimony of cells themselves. Just as scholars who deciphered the hieroglyphics of ancient Egypt, the cuneiform symbols of Mesopotamia, and the Linear B script of Mycenaean Greece enriched our understanding of these extraordinary civilizations, biologists struggling to decipher the chemical language of cells have opened another window on the civilization of multicellular organisms. In the conversations between cells, they have discerned the details of everyday life in cellular society. In addition, they have observed firsthand how miscommunication between cells can precipitate a catastrophe. Too much or too little of a signaling molecule, a gene mutation that compromises a cell's hearing, a defect that blocks the free flow of information within the cell, outside interference from foreign chemicals can all lead to confusion instead of communication. Cellular language researchers now know that such pathological misunderstandings lie at the heart of some of our most intractable diseases: cancer, diabetes, obesity, addiction, autoimmune disorders. Slowly, haltingly at first, and then with increasing confidence, their understanding has enabled them to speak to cells as well, fashioning drugs to talk sense to our bodies when degeneration and disease create confusion or misunderstanding.

For nearly a century these researchers have eavesdropped on the conversations between cells, hoping to master the basics of their language. This book is the story of their discoveries as well as an anthology of the tales they've heard cells tell. It describes how cell-to-cell communication shapes a fertilized egg into a body, maintains tissues, regulates the distribution of resources, records memories, and builds a firewall against invaders. It explains how the disruption of signaling pathways can precipitate communication breakdowns that result in well-known medical conditions. Finally, it reveals how farsighted researchers are beginning to apply what's been learned about cellular communication to the solution of biology's ultimate problem: how the processes that sustain life emerge from the collaborations between inanimate molecules. In doing so they hope to lead the way out of reductionism toward an integrated biology that acknowledges the complexity of living organisms.

Children master languages with ease, but for adults, learning to speak any new language is difficult. The words are strange and may be hard to pronounce. The rules of grammar are unfamiliar. The culture itself, its idioms and customs, baffle and conspire to embarrass. Until a student has acquired a rudimentary vocabulary and learned the correct way to join those words into sentences, even a simple conversation can be a struggle. The language spoken by cells is no different. Though I have endeavored to make the unknown less formidable by streamlining sentences and reducing the use of acronyms and abbreviations, you may feel at first as if you're back in high school, trying to conjugate foreign verbs. Be patient; practice makes perfect.

The language of cells is the language of modern biology. Because a working knowledge of cellular signaling mechanisms "is essential to our understanding of the control of virtually all biological processes," according to the editors of the journal *Science*, cell communication has become one of the hottest topics in biomedical research. Over the past several decades, the number of scientific papers published on cellular communication has skyrocketed, not only in journals read by cell biologists but also in those devoted to neuroscience, immunology, pharmacology, physiology, developmental biology, infectious diseases, and molecular biology as well as clinical research journals. What I present here is, by necessity, only an overview of this vast literature. I apologize to any in the scientific community who feel I have omitted or glossed over their favorite topics and offer in my defense the hope that readers will be left wanting to know more and will have acquired the language skills they need to satisfy this curiosity.

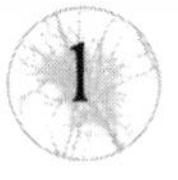

SMALL TALK

In broad daylight, you are floating in darkness. You swim in silence because you cannot hear. On your right, a cloud of sugar molecules drifts just a few centimeters away—dinner, but how will you find it? On your left, a noxious chemical seeps toward you—but unless you know that you're in danger, how can you escape? If you are the bacterium *Escherichia coli*, sharing space in the human gut with other indigenous flora and fauna, this is your world: unpredictable, at times even inhospitable. But don't dwell on your limitations. Your kind came to life when the earth was still hot and sulfurous and has used the intervening eons to craft all the tools and techniques you need to survive in a capricious environment.

You have many talents. For one thing, you can move. Let the bottom dwellers and the stone huggers ache to be noticed by some passing current; evolution taught you to waltz—chassé, pause, spin, glide, ONE-two-three, ONE-two-three, left, right, zigzag. Your dancing shoes are a skein of protein filaments, or flagella, powered not by muscles and tendons but by a gearbox of proteins that operate as a rotor, twirling the flagella at more than 100 revolutions per second. When the rotor turns counterclockwise, the flagella spiral into a single tail and you glide in a smooth, straight line. When it spins clockwise, the flagella unfurl and stroke, each to its own beat, and you spin and tumble in place. Reverse again and you resume swimming in a new direction. The rotor switches back and forth as regularly as a metronome, spinning one way for a few seconds, then the other. Spinning and swimming, you meander along to its rhythm, improvising a daydream of a dance that textbooks refer to as the "random walk."

It's a good thing you don't have a mother—she'd surely admonish you to watch where you're going. You can't afford to be so oblivious, she'd scold, or you're liable to waltz right into trouble. As neurobiologist Rodolfo Llinas warns human ramblers, "Active movement is dangerous in the absence of an internal plan subject to sensory modulation. Try walking any distance, even in a well-protected, uncluttered hallway, with your eyes closed. How far can you go before opening your eyes becomes irresistible?" But you could tell her not to worry—you can chart a less haphazard course when you need to. Should an appetizing snack appear on the horizon, the rotor lingers in counterclockwise mode, so that you swim straight ahead, in the direction of the meal, instead of meandering aimlessly. A distasteful substance, on the other hand, shifts the rotor clockwise, and you tumble in search of an escape route—the path you'll follow when it resumes its usual alternating pattern.

By repeatedly adjusting the proportion of clockwise-to-counterclockwise rotation, you can forego your wandering ways in favor of a one-step forward, one-step-sideways shuffle—not exactly what a more advanced creature might think of as purposeful movement but a "biased random walk" that hitches determinedly, if somewhat erratically, toward satisfaction or away from catastrophe. Microbiologist Ann Stock explains: "If the cell finds itself moving in the proper direction, it suppresses tumbling and moves further in that direction. Then it randomly reorients and heads off in a new direction, one that might still be good or may now be bad. If it's going in the

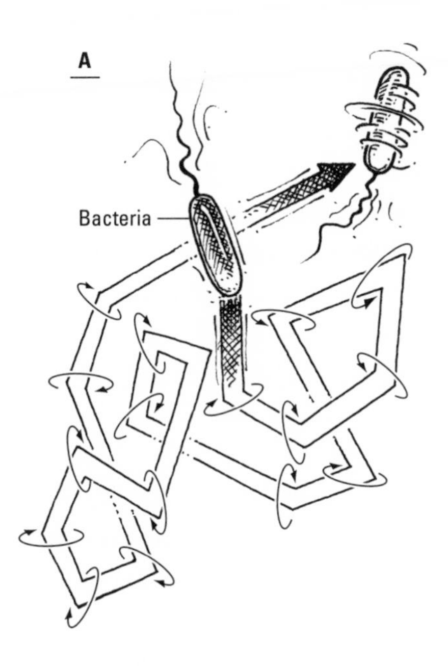

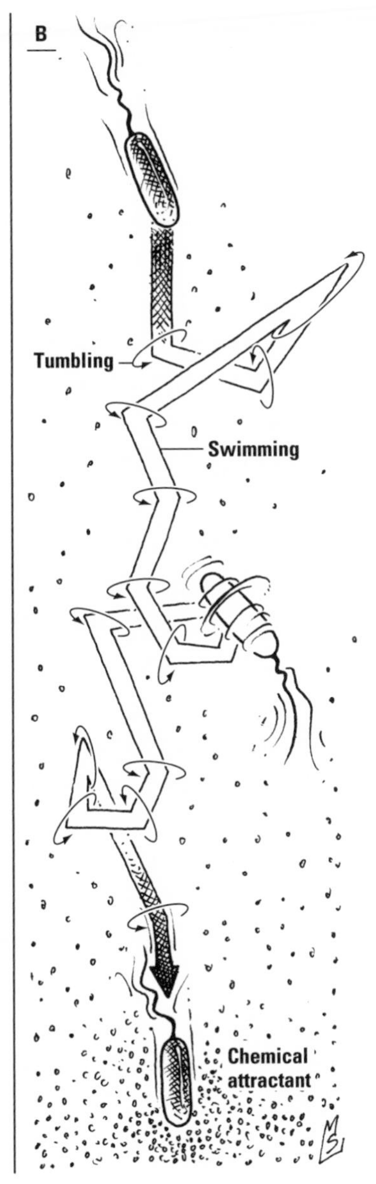

E. coli, out for a walk. In the absence of an attractant (an edible amino acid, oxygen) or a repellant (distasteful metal ions, for example) stimulus, the bacterium alternates randomly between periods of tumbling in place and smooth swimming by switching the direction in which its flagella rotate (*A*). Should *E. coli* detect a chemical signal, however, the information will be relayed to the motor turning the flagella, causing the flagella to rotate preferentially in one direction. Here (*B*), an attractant has encouraged rotation in the counterclockwise direction; as a result, the bacterium swims more than it tumbles, moving in the direction of the food source.

wrong direction, it goes back to tumbling. It seems tortuous, but the same patterns are used by grazing animals to find better pastures. You do a little sampling in all directions and keep going when things are good and reorient when they're not." But how did you determine which direction was the "right direction"? And how did you use that information to change the rotation of your flagella?

I look and listen and touch; if I could swim with you, I would know the human gut as wet and warm, a muffled rush of murky water. But there are other ways to experience life. Your world is vivid with chemicals as well as colors, molecules as well as sounds. They could be your connection to the outside world, the cues that could chart your path. Even this solution presents difficulties, however. Like other cells, you are surrounded by a protective membrane of lipids and proteins isolating your fastidiously composed internal fluids from the unpredictable excesses of your surroundings. Nothing that might upset the delicate balance critical to life can penetrate this barrier—but neither can the chemical signals bearing news about current events. Inside your hermetically sealed bubble, the rotor proteins controlling your flagella are waiting for direction; outside, messengers with the information needed by the rotor proteins mill restlessly in front of closed doors.

Troublesome organism, isn't there any end to your problems? Evolution should have just given up on you, allowed you to starve in silence. Instead it has crafted words to describe your world and rules for combining them, the building blocks of a language that tells you how to move in step with the world around you and gives your crooked walk a glamorous name: "chemotaxis," movement directed by chemicals.

BETTER MOVEMENT THROUGH CHEMISTRY

The ancestors of *E. coli* did not have plastics, but they did have another type of polymer suited to a wide range of applications: proteins. More heterogeneous than polystyrene or polyurethane, nature's megamolecules are combinations of 20 different building blocks, or amino acids, coupled to each other by the electron-sharing arrangement known as a peptide bond. Man-made plastics can be shaped, molded, or extruded to order, but protein polymers automatically bend themselves into useful shapes, guided only by their affinity for water and the social relationships between their constituent amino acids—no heat or specialized machinery required.* Scores of fragile alliances between amino acids act in concert to stabilize the folded protein, resulting in a three-dimensional configuration as unique as a fingerprint.

Productive only when wedded to another molecule, the protein relies on this structure to play the role of matchmaker, embedding an advertisement for a soul mate in the loops, bulges, and trenches created on its outer surface by the folds:

> "**Are You My Better Half?**"—SFP (single folded protein) with secure position in healthy cell seeks compatible molecule with interest in chemical engineering, architecture, or communication for exclusive short-term relationship. Please reply with details of your chemical structure.

The molecule with the most compatible profile wins the date and enters into a marriage of convenience, arranged by evolution to accomplish a specialized task. Paired with a matching substrate, an enzyme speeds up a vital chemical reaction. A rotor protein and a flagellar protein conspire to move a bacterium. And a protein with a soft spot for a nutrient or a toxin (call the partners "receptor" and "ligand"), as well as a torso rich in water-repelling, or hydrophobic, amino acids on intimate terms with the lipid-rich plasma membrane, is the perfect spy, relentlessly drawn to important information and

*Many proteins do, however, need an assistant—a second protein known, appropriately enough, as a "chaperone"—to complete the job of folding.

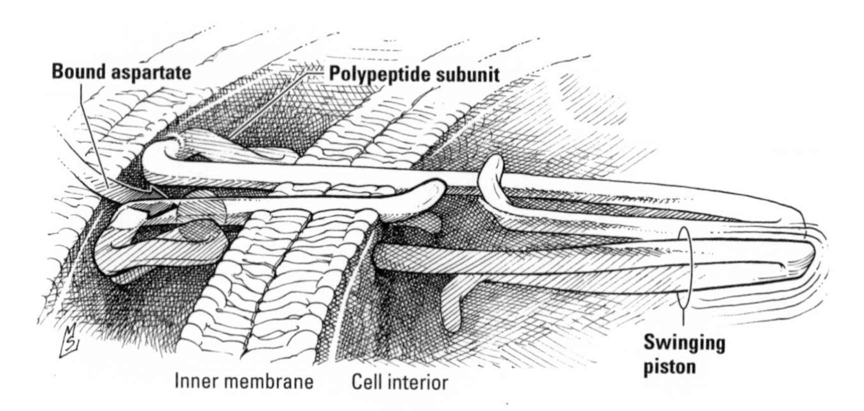

The Tar receptor. The binding site of this sensor protein, which can detect both the amino acid aspartate and repellants such as nickel, projects above the outer surface of the plasma membrane. A mobile helical segment spanning the membrane acts as a "swinging piston" that nudges the first member of an intracellular team of proteins charged with relaying the news to the flagella.

built to share that information with an insider able to put it to good use.

E. coli sponsors five such receptor-ligand partnerships—its five senses. The receptor that biologists call "Tsr" is married to the amino acid serine. The Tar receptor's partner is aspartate. Tap prefers peptides; Trg falls for proteins that bind and transport the sugars ribose and galactose. Aer is devoted to oxygen instead of nutrients. To conserve precious space in a small genome, adultery is not only permissible but encouraged; Tar, for example, has leave to consort with repellants, such as nickel ions, in addition to aspartate.

The architecture of the receptor determines how it will speak as well as to whom it will speak. Look, for example, at the Tar receptor in cross section (*you* can't, of course, but X rays or the magnetic field generated by a nuclear magnetic resonance spectrometer can), and you'll see that it's not a single protein but a dimer (pronounced "DIE-mer")—a pair of identical polypeptide subunits. Note how each rears up out of the membrane like a flower, crowned with a rosette of four helical segments that cradle Tar's preferred ligands. Two of these helices wind out of the binding site and through the membrane, tapping into the rich cytoplasm below. Rooted, Tar is not rigid, however. During the subtle shape-shifting triggered when receptor and ligand embrace, one transmembrane helix is pushed downward and forward. The kick delivered by this "swinging piston" transmits news of the binding of attractant or repellant from outside to inside; in effect, Tar uses its own body to announce the discovery of ligand, much as an electronic sensor might activate a buzzer or a flashing light.

In an organism as small and simple as *E. coli,* one word (or one nudge from a dangling helix) should be enough to tell a rotor "left" or "right." But this bacterium must have had a finicky language

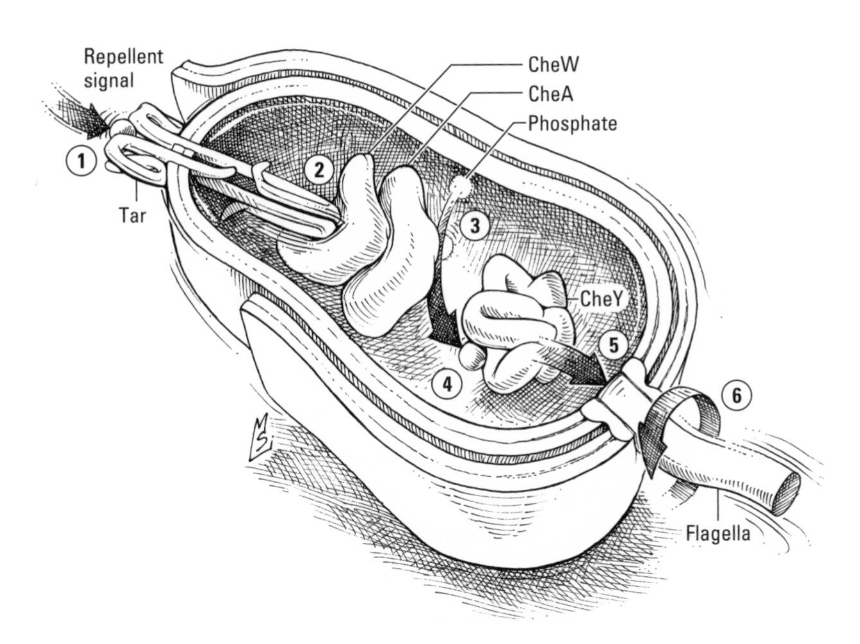

A bacterial "sentence." The binding of a repellant signal to the Tar receptor (*1*) sets the helical piston swinging (*2*), prompting CheW, the "word" heading the intracellular portion of the sentence, to activate the kinase CheA (*3*). This enzyme transfers a phosphate group to CheY (*4*); phosphorylated CheY (*5*), in turn, tells the rotor protein, currently turning counterclockwise (*6*), to change direction.

teacher, for it insists on speaking in complete sentences. The receptor is the subject of this sentence, the predicate a team of intracellular molecules called "Che" (for *che*motaxis) proteins that relay the message from receptor to rotor and add the nuance of context. CheW, the team member that heads this bucket brigade, is merely a go-between. When Tar detects a repellant, CheW takes the kick, then pokes the next protein in line, CheA, the action verb at the core of every chemotactic sentence.

One of a large and influential family of enzymes known as "kinases" (from a Greek word meaning "to move"), CheA, like its kin, responds to a wake-up call with an expletive—"Phosphate!" The chemical group that puts the "P" in ATP, phosphate, bristling with three negative charges, has the power to twist proteins inside out, turn ordinary amino acids into powerful magnets, and disrupt long-standing relationships—a switch flipping switches. "Nickel! Pass it on," whispers CheW, and the excited kinase spits a phosphate group into the face of one of its own histidine residues. The histidine, in turn, spews it at the nearest bystander, an aspartate protruding from the final protein in the relay, CheY. To CheY, phosphate is not an insult but the gift of wings. It soars across the cell, landing on the rotor protein. "No more swimming!" it bellows. The rotor obediently shifts into clockwise mode. Flagella flail, and the bacterium spins like a child on a tilt-a-whirl.

The news of aspartate binding outside the cell travels via the same protein relay but has the opposite effect. Instead of agitating, the receptor tells CheW to tell CheA to keep its mouth shut. The kinase dozes; CheY stays home and keeps to itself. And without phosphorylated CheY meddling with the rotor, the bacterium continues to swim, gliding smoothly toward the food source.

Writers are admonished to be sparing in their use of adjectives, but *E. coli* shamelessly adds two modifiers to every chemotactic sentence. Such verbosity isn't a sin in this language, however—the extra words, the enzymes CheR and CheB, add contextual information the bacterium needs to fine-tune responses, in the form of methyl groups (a "methyl group" is a chemical "special teams" unit consisting of a carbon atom surrounded by three hydrogen atoms) that they plug in to or pull out of sockets in the tails of chemotactic receptors. Whenever the binding of an attractant stills the receptor and quiets CheA, CheR inserts methyl groups into the sockets; with each addition, the receptor recovers more of its voice. The binding of a repellant, on the other hand, prompts CheA to activate the CheB enzyme along with CheY. Phosphorylated CheB unplugs methyl groups, hushing both receptor and kinase. By adding and subtracting methyl groups, the two enzymes can adjust receptor sensitivity to mitigate the effect of ligand binding. "You can think of chemotaxis receptors as sitting in a balance between the ligand binding state and the methylation state (that is, the number of methyl groups attached to the receptor protein)," explains Ann Stock. "When the two are properly balanced, you have a steady state output. If you add a stimulus, you throw the receptor into a new signaling state, where you get some sort of response. Then, if you change the methylation state to counterbalance the presence of ligand, you restore the initial steady state output." This balancing act, known as *adaptation*—the bacterial equivalent of the adjustment your eyes make when you walk out of a dark room into bright sunlight—links the response to a change in the amount of ligand, rather than concentration per se, extending the range of the chemotaxis system over five orders of magnitude.

Methylation not only keeps bacteria alert but also makes them smarter; in addition to volume control, it serves as a primitive form of memory. A physical record of the bacterium's most recent interaction with the environment, the methylation state of the receptor records the concentration of attractant or repellant. Then, a few strokes later, the organism can reference this value, compare it to the current level of stimulation, and engineer a midcourse correction if necessary. Stock concludes: "The cell is testing out its environment at one point in time. It's swimming, it's moving to a new place, comparing. And then it makes the decision whether to go forward or tumble based on what has happened over time."

"The modern era begins, characteristically, with a revolution," writes historian Jacques Barzun in *From Dawn to Decadence: 500 Years of Western Cultural Life*, referring to the cultural upheaval that began with Martin Luther's challenge to the Catholic Church. However, as Barzun notes, this cataclysm did not simply erupt out of the blue. The influence of earlier reformers and critics, the invention of the printing press, the introduction of higher-quality paper and ink, and the emergence of skilled printers provided the ideas and tools necessary to transform a local protest into an international spectacle.

Come here. Run away. Just as the Protestant reformation had its roots in the earlier efforts of Wycliff and Gutenberg, the revolution in communication that would one day guide interactions between cells almost certainly had its origins in simple directives like these, governing interactions with the environment. Pioneered by a distant ancestor fortunate enough to stumble upon the virtues of proteins for engineering alliances between molecules, the chemotactic sensory apparatus of bacteria like *E. coli* illustrates fundamental principles common to the transfer of any type of biological information via chemical signals: the exploitation of protein topography to discriminate signals; the use of transmembrane receptors to circumvent the barrier posed by the plasma membrane; the coordination of perception and response, as well as the integration of multiple signals, by means of protein relays featuring kinases; the regulation of receptor sensitivity by chemical modification. Using these principles as a template, evolution would construct a chemical language that would end the isolation of cells, that would allow them to talk to one another, to cooperate, to live as a group yet behave as a single organism—one of the most extraordinary achievements in the history of life.

GROUP DECISIONS

"You want a crab?" A skinny boy of about 10 holds a saucer-si crustacean by a front claw and shakes it at me.

"That's a big one. You're not taking it home?"

His mother, who is trying to persuade his younger brother to pick up the sand castle paraphernalia, looks up and glares at me. "It already smells," she hisses. Behind her, two hungry seagulls nod in agreement.

"Sure," I reply, and hold out a small plastic bucket.

Back at our beach blanket, my 13-year-old daughter Haley looks up from the latest issue of *Teen People* and peers into the pail. "Why do you have a dead crab?"

"It's not dead; it's just not moving." I prod gingerly at the crab with the end of a pencil, but it has turned peevish and uncooperative. It squats at the bottom of the pail like a sullen garden toad, refusing to budge. I poke more aggressively. Without warning the stalks supporting the crab's beady black eyes shoot straight up at me. "Ahhh! Did you see that? It popped its eyes at me!"

"It's being weird because it's in a bucket of hot water," Haley remarks, turning the page.

"I was going to draw its picture, but I think maybe I'll just let it go. Would you like to help me?" I ask.

We pick a spot near a rocky outcrop, checking first for the gulls. When I'm certain they've moved on in search of other prey, I tip the pail into the surf. Torpid in captivity, the crab dances in the saltwater, energetic as a wind-up toy. It punches at the foaming wavelets, scrambles along the edge of the rocks, cartwheels, hops. It clutches at the slippery rocks. Finally, it darts sideways into a crevasse and disappears, safe at last from hungry gulls and grasping humans.

Dozens of species of marine bacteria also call this tidal flat home—and they don't take any more kindly to deportation than my crab. Relocated to the laboratory, most die, no matter how attentively they're coddled. Microbiologists at Northeastern University suspected this failure to thrive was not the result of bumps and bruises incurred during the move itself but of homesickness afterward; to prove it, the researchers decided to culture the beach along with the bacteria. They cut out a block of beach sand and isolated the resident bacteria. Instead of transferring the microbes to petri dishes, however, they corralled them in small diffusion chambers formed by sealing a washer between two permeable membranes, placed them back on their native sand in a marine aquarium, and submerged the chambers in seawater, in effect "returning" the bacteria to the wild. Cultured the traditional way, less than 1 percent of the bacteria originally isolated would have survived; surrounded by the comforts of home, 22 percent, including two previously unknown species, not only survived but thrived.

In the laboratory incubator, it is never winter—but it's not home either. "We've tended to use bacteria as model systems in the laboratory," says Ann Stock. "But we realize now that the conditions that you keep them under in a laboratory don't even come close to scratching the surface of the kinds of environments they find themselves in when they're living in the real world." For some species, like these marine microbes, the difference between their native environment and the lab environment is more than their fragile physiology can handle. Others are more resilient; a few, like *E. coli*, adapt so well, Stock notes, that it's easy to forget they were once wild creatures. Unless culture conditions demand some effort, these domesticated bacteria find it easy to forget survival skills that were essential in the wild; after all, why waste effort on rotors and receptors when safety is a given and food is delivered regularly to your doorstep? If you don't use it, it's best to lose it. "There's a tremendous selection for efficiency," Stock observes. "We see all the time in the laboratory that cells will lose genes or the expression of genes and behaviors that you'd see in the wild-type environment because you don't maintain selection for them."

Shielded from the demands of the real world, many domestic

bacteria have not only lost their ambition but also mothballed a talent that until recently no one knew they even had—the ability to talk. Speechless in agar, bacteria at liberty in soil and pond water, milk and meat, the open sea, or the human body are not mute, but chatter relentlessly to each other, in chemical dialects that follow the same organizational principles as chemotaxis. They trade gossip. They sound alarms and plan invasions. And, most surprising of all, they form alliances, primitive communities that allow them to enjoy benefits—more efficient use of resources, access to environmental niches unsuited to single cells, safety from predators—once thought to be the sole province of "traditional" multicellular organisms like plants and animals.

"Certainly bacteria are the best-studied organisms," says microbiologist Bonnie Bassler. "But they have always been considered to live these sorts of individualistic, asocial lives. And it's just not true." In their native environment, bacteria are social creatures, eager to collaborate and able to do so because, like more complex organisms, they have discovered that the key to civilization is communication.

Unlike *E. coli*, the roses beautifying my fence don't have to watch where they're going because they aren't going anywhere. Plants, able to feed themselves by trapping sunlight, have sprouted roots instead of flagella and thrown in their lot with the soil. Staying put is a good way to avoid running into trouble but no way to meet members of the opposite sex. Enter the honeybee. In return for a contribution to its pantry—a share of the pollen and sweet nectar to be made into honey—it does the heavy lifting of reproduction for the rose, distributing pollen like love letters as it wanders from flower to flower. It's a win-win proposition for everyone: the plant finds a mate; the bee feeds its brood.

In the moonlit shallows off the coast of a South Sea island, the Hawaiian bobtail squid *Euprymna scolopes* and the marine bacterium *Vibrio fischeri* have negotiated an equally beneficial arrangement.

Night is a dangerous time to trawl these waters. The squid's shadow ought to give it away; instead, all a hungry fish sees is silver—*Euprymna*, bright as the moon, is invisible to its enemies, thanks to the hospitality it extends toward *V. fischeri*. In return for room and board in a sac on the underside of the squid's body, *V. fischeri* produces an enzyme, luciferase, facilitating a chemical reaction that releases energy in the form of light. The luminescent bacteria ensconced in the light organ enjoy a free meal, while the illuminated squid avoids becoming someone else's.

Keeping the lights on all of the time is expensive, whether you're paying for kilowatt-hours or burning food to synthesize light-making enzymes. When it's on its own in the open ocean, therefore, *V. fischeri* hides its lights under a bushel, switching off the luciferase gene until it's inside the light organ of a cooperative squid. The bacteria "know" when they've reached this safe haven the same way teenage girls know when they're at the "right" party—all of their friends are there. *V. fischeri*, of course, can't take a head count by cell phone. But it can secrete chemical messengers to advertise its presence, and it can listen for signals secreted by others, drawing on the same principles underlying simple sensory mechanisms for keeping track of the physical environment to develop similar mechanisms for keeping track of the social environment.

In the language of *V. fischeri*, the word for light is "*N*-(3-oxohexanoyl)-homoserine lactone." Non-native speakers are welcome to use the colloquial term "autoinducer," coined by Harvard researchers Kenneth Nealson and John Woodland Hastings, when they proposed, over 30 years ago, that a chemical substance synthesized and secreted by the bacteria themselves cued the expression of the luciferase gene. Even that's a large word for a molecule so small the plasma membrane poses no barrier to it—it leaks out of the bacterial cell almost as fast as it's made—and it's being made most of the time, because *V. fischeri* has a habit of muttering to itself more or less constantly, even when no one's listening. In open water, where individual cells are rarely in close contact, the signal is diluted so much that by the time it reaches the nearest neighbor it's inaudible. In the cramped quarters of the squid light organ, however, where as many as 10 billion bacteria crowd together, concentrations of autoinducer swell to cacophonous proportions, sending an unmistakable message to every member of the group: "You are not alone."

The chemotaxis signaling mechanism employs a relay of proteins, but *V. fischeri* is terser; the sudden flood of autoinducer molecules activates a dormant protein, LuxR, that is both receptor and response regulator. CheY is only a lowly mechanic, fiddling with a rotor. LuxR is an artist, the genome its medium. One of an elite class of proteins known as transcription factors, it has permission to turn genes on and off: luciferase—on; *luxI*, which encodes the enzyme responsible for synthesizing the autoinducer—on; *luxR*, the receptor gene itself—off. The production of signal and light, increased by LuxR, is simultaneously tempered by a compensatory decrease in receptor number, allowing *V. fischeri* to fine-tune the light response, just as the addition and subtraction of methyl groups allow *E. coli* to adjust the sensitivity of chemotaxis receptors.

A single bacterium can generate only a pinpoint of light. But a group of bacteria, operating as a team, can produce enough light to forge a mutual defense pact with a larger organism. "It's mob psychology," quips Bonnie Bassler, but it's also a model of parliamentary procedure: all decisions are deferred until a requisite number of voters can be assembled. "If the bacteria wait, and they count themselves, and then they all do it together, they can have an enormous impact and frankly overcome tremendous odds," she continues. Known to microbiologists as "quorum sensing," this show of hands allows the group to "organize [itself] to act like a huge multicellular organism," reaping the attendant advantages of cooperative living as a consequence.

The recipient of a 2002 "genius" award from the MacArthur Foundation, Bassler began working on quorum sensing a decade ago,

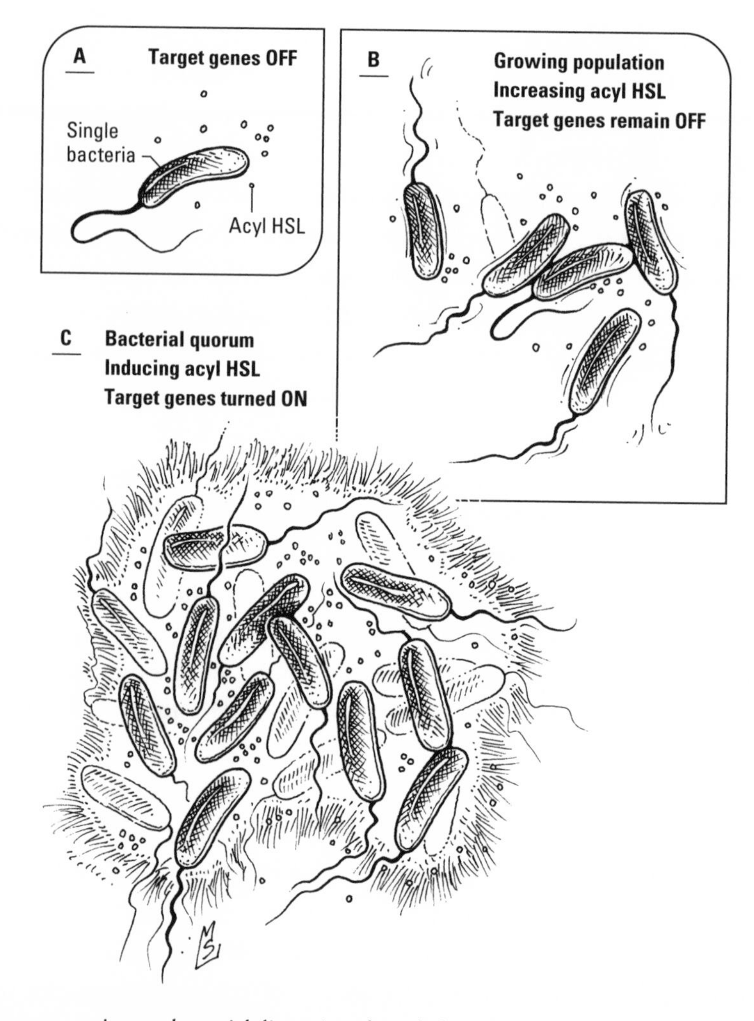

Quorum sensing—a bacterial discussion about light. The bacterium *V. fischeri* emits a chemical signal, known as an acyl HSL, to announce its presence to others. In the open sea this signal rapidly dissipates (*A*). Within the confines of the squid light organ, however, the concentration of acyl HSL increases as the number of bacteria increases (*B*). When the concentration reaches a threshold level, the signal triggers the expression of genes encoding the proteins responsible for the light-generating chemical reaction (*C*).

when most biologists thought of *V. fischeri* as a freak of nature, its discussions of population density no more than "an odd anomaly of these interesting but not very important glow-in-the-dark bacteria from the sea. It was considered fringe science," she recalls. Today, quorum sensing is hot gossip, thanks to research by Bassler and others demonstrating that *V. fischeri* isn't the only species counting with chemicals. "All bacteria probably do what it does," she maintains. "And they don't just turn on light; they turn on all kinds of functions that make sense for bacteria to express when they're in a community but not when they're alone. There's an enormous range of behaviors besides luminescence that are regulated by cell-cell communication."

More than 50 other species of Gram-negative bacteria (so named because of their reaction to a commonly used staining procedure) count out loud by releasing chemicals similar to *V. fischeri*'s autoinducer, known collectively as acyl homoserine lactones—acyl-HSLs for short. In nearly every instance, a LuxI-like enzyme synthesizes the acyl-HSL signal, which is detected and translated into action by a LuxR-type receptor/transcription factor controlling critical genes. But as Bassler says, not everyone is interested in discussing how to make the world a brighter place. Many—take *Pseudomonas aeruginosa*, for example—are busy plotting murder and mayhem. This deviant, which preys on the injured and the immunocompromised, brings weapons instead of light when it comes to visit: so-called virulence genes, encoding proteins that allow it to stick to, penetrate, and poison host cells. But if *P. aeruginosa* launches its attack before the expeditionary force has multiplied into an army, it risks tipping off the victim's immune system. Communication by quorum sensing helps the bacteria time the release of their virulence factors to coincide with a critical population density, evading detection until the cohort is large enough to take on the defenders and win.

Gram-positive bacteria, such as the common soil microbe *Bacillus subtilis,* also count heads. In the dark, damp earth of the forest floor, *B. subtilis* relies on quorum sensing, in the context of current economic conditions, to choose between two retirement options. In well-to-do communities, the quorum opts to stay put and maintain the family home, promoting each other to "competent" status, in which everyone can activate genes for the machinery needed to scavenge DNA fragments from dead colleagues and repair defects in their own genomes. Bacteria confronted with a dwindling food supply, on the other hand, call for hibernation and migration. Neighbor croons to neighbor a chemical lullaby that puts everyone to sleep as spores, until fresh provisions arrive or a passing creature, a breeze, a storm-instigated rivulet sweeps them off to greener pastures.

Acyl-HSL isn't spoken here, however. *B. subtilis* and other Gram-positive bacteria have their own way of saying things, in oligopeptides, short chains of amino acids cut from larger proteins and shipped out of the transmitting cell on the back of a dedicated transport protein. Instead of an all-in-one receptor/response regulator, Gram-positive bacteria are equipped with a "two-component" signaling relay, consisting of a receptor that is also a histidine kinase and a second protein that is a transcription factor. Well-fed bacteria say "Com X," and the binding of oligopeptide and receptor leads to the phosphorylation of the transcription factor, enabling it to activate competence genes. Starving bacteria call out "CSF," and everyone curls up into a spore. Skillful communication promotes sensible decisions; in times of plenty, the bacteria maintain the integrity of their genomes instead of languishing as spores, while in the face of imminent catastrophe, they postpone rebuilding and hunker down to weather the crisis.

Life may have begun as a single cell, but it did not stay that way for long. "Several hundred millennia ago, prehistoric humans learned that there is strength in numbers," Bassler and colleague Stephen Winans note in a recent review, but "bacteria made this discovery at least a billion years earlier," a discovery destined to preside over the birth of cooperative behavior and social organization. Just as "the evolution of language likely aided in the ability of protohumans to coordinate the behavior of the group," the evolution of chemical signaling—quorum sensing—facilitated the emergence of group behavior in bacteria. As a result, they argue, quorum sensing is not simply a bacterial idiosyncrasy but "one of the first steps in the development of multicellular organisms."

THE TALK OF THE TOWN

"Now the whole earth had one language and few words." According to the biblical account of the Tower of Babel, the human race wouldn't need translators if it weren't for a few arrogant architects—we'd all still be speaking the same language. Instead, a bewildering array of at least 6,000 different tongues and regional dialects confounds international communication, impedes commerce, and undermines diplomacy. No wonder enterprising wordsmiths have tried to create a "universal language" that would allow speakers of English and German, Pashto and Hindi, Swahili and Tagalog to talk freely with one another. One of the best known—and arguably the most successful—of these artificial languages is Esperanto (meaning "one who hopes"), created in 1887 by a Polish doctor, Ludwig L. Zamenhof. An amalgamation of Latin, Romance languages, German, English, Greek, and Slavic, Esperanto features a simple, law-abiding structure devoid of the irregular verbs, gendered nouns, and arcane spellings that plague natural languages; as a result, aficionados claim, it's five times easier for a native English speaker to learn than French or Spanish, 10 times easier than Russian, and 20 times easier than Chinese.

Perhaps bacteria, unable to build towers that offend deities, have been rewarded for their humility. Perhaps they have simply been prudent closet cleaners, careful as they updated and revised their genomes over millions of years, not to lose their universal language, chemical signals understood by all, in addition to the secret dialects each species uses to talk to its own kind. Despite its carefully engineered simplicity, Zamenhof's Esperanto can count perhaps 2 million speakers worldwide, but this "bacterial Esperanto," as Bonnie Bassler calls it, may boast millions of speakers in a square centimeter, all of whom were born knowing it. Invaluable in a world teeming with strangers, these universal signals enable bacteria to ask not only "How many?" but also "Who are you?"

If you want to be a molecular geneticist, why not make your life easy by studying an organism that's small, economical, and doesn't bite, that has only one chromosome, and that spawns a new generation every 20 minutes? And if you want to make the job even easier, why not go one step further and select a microorganism that has a heritable trait you can study with the naked eye simply by turning out the lights? That's why Bonnie Bassler chose *Vibrio harveyi*.

Bassler's affection for seafaring bacteria began in graduate school, when they rescued her thesis. "I was working on a project that involved chicken cells and it was going nowhere. Then my thesis advisor got a small grant from the Navy to study how bacteria adhered, because when boats get barnacles on the bottom, the first thing that happens is you get bacteria sticking to the surface. So he handed me a little vial of marine bacteria and told me to figure out how they stick," she recalls. Afterward, when she wanted to master the tools and techniques of the DNA trade, the incandescent talents of marine *Vibrio* species again made them the perfect lab partners. "As a wannabe geneticist, if you're going to make mutants, all you have to do is turn out the lights in the room and see if they don't make light or if they glow when they shouldn't. This is perfect for me because what could be easier?"

Bassler came to the Agouron Institute in La Jolla, California, to work with Mike Silverman, the "one person who was doing genetics in these wild bacteria" as well as the pyrotechnician who had spearheaded the effort to identify the components of *V. fischeri's* LuxI/

LuxR quorum-sensing mechanism. Under his direction she used a similar approach to dissect *V. harveyi's* quorum-sensing system and discovered that it had a different way of talking about light. In contrast to *V. fischeri*, this Gram-negative bacterium ferried messages to genes via a two-component relay like that found in Gram-positive species. And it secreted a second autoinducer—but this one, christened simply "AI-2," wasn't an acyl HSL. What sort of molecule it was, no one could say, because AI-2 resisted purification and defied every attempt to determine its chemical structure. Scientists may have been mystified. Among bacteria, however, Bassler discovered that AI-2 was common knowledge.

Now with her own lab at Princeton University, Bassler and a new team of co-workers cloned the gene responsible for producing AI-2, and, as is common practice, compared their sequence to that of other bacterial genes on file in public databases. To their surprise, the search uncovered a nearly identical gene in more than 40 other species of bacteria, Gram-positive as well as Gram-negative, including a "list of the clinical 'Who's Who' in pathogenesis": *Vibrio cholerae*, the cholera-causing black sheep of the *Vibrio* family; several *Salmonella* and *Streptococcus* species; the respiratory pathogen *Haemophilus influenzae*; *Borrelia burgdorferi*, responsible for Lyme disease; *Helicobacter pylori*, maker of ulcers; and *Yersinia pestis*, infamous as the causative agent in bubonic plague. "As far as we can tell," says Bassler, "they're all making the identical molecule. In the acyl-HSLs, the acyl side chain differs from bacterium to bacterium, and that confers exquisite species specificity. We think this is more of a universal language." In fact, she suggests, AI-2 may be one of the oldest and most familiar words on earth. Because it's spoken by both Gram-negative and Gram-positive bacteria, it is likely to have evolved before the two lineages diverged—and before they began speaking in acyl-HSLs or oligopeptides.

Not only did Bassler determine that AI-2 is a universal signal, she finally solved the mystery of its chemical composition. She knew that the synthesis of AI-2 began with a compound that cells often turn to when they need methyl groups, *S*-adenosylmethionine. And she had learned that an enzyme, LuxS, finished the job, but wasn't a meticulous chemist. The floppy carbon skeleton that was LuxS's handiwork could be folded in several different ways. So-called pure preparations were actually a mixture of all these configurations, only one of which was the structure preferred by living bacteria. "It was like sludge," says Bassler. Then she came up with the idea of using the AI-2 receptor—which could be purified—to trap the correctly folded AI-2 signaling molecule. "We wondered, 'Couldn't we take the receptor to go into this mix and pick out the right structure?' So my colleague and friend Fred Hughson crystallized the sensor protein bound to the true autoinducer. That's how we got the picture." The "picture"—the chemical structure of AI-2, as revealed by X-ray crystallography and mass spectrometry—resembled a pair of pentagons glued long end to long end and riveted together at the top by an unusual bolt: a boron atom. This was the quirk that had foiled so many previous attempts to identify AI-2. "Boron has had this mystery role in biology," explains Bassler. "It's known that boron is required by all these different organisms, but no one knew any reaction that it was involved in until this one. It's like brand-new chemistry."

Examination of the synthetic path from *S*-adenosylmethionine to AI-2 also suggested why this particular molecule might have been available to so many kinds of bacteria: its synthesis provided a solution to the universal problem of what to do about *S*-adenosylhomocysteine, the toxic metabolite left behind after *S*-adenosylmethionine relinquished its methyl group. "It's a detoxification pathway," Bassler notes. "The bacteria almost have to make it, because of this toxic intermediate." In the process they inadvertently discovered a way to fashion a molecule with a message: I am.

Such self-aggrandizement led to grander things than counting family members. Bacteria that could speak in words understood by all could build relationships with other species, other cultures. They could build consensus. They could even build cities.

It's a footloose existence, swimming and tumbling in a flask, through the gut, across a pond. But most bacteria are not nomads at heart; given a choice, what they really want is a place to call home, a refuge from the tiresome rush of air or water, where they can cling and congregate. A rock, a pipe, your teeth, plastic or glass, even the interface between a container of water and the overlying air, are potential building sites—provided they're up to standard, because bacteria are as choosy about the neighborhood as any other potential homeowner. For some species a feast of sugars, or the perfect blend of amino acids, is the selling point, as compelling as an ocean view or award-winning schools. Some want fresh air, oxygen; some are drawn like a microscopic magnet to iron. Others need balmy temperatures or a particular salt concentration. How they would annoy a real estate agent, inspecting and rejecting neighborhood after neighborhood—warm, cold, luxurious, Spartan, salty, metallic, well aerated—in search of the illusive site that meets all of their demands.

"This is not a step they take lightly," insists microbiologist Roberto Kolter. "Bacteria don't just run into a surface. It's a decision they make, to go and assemble themselves on that surface." In his office at the Harvard Medical School, Kolter touches a key on his computer keyboard to start a video documentary he's made that traces the natural history of one such settlement. Black dashes dart across the screen, like grains of wild rice shaken on a sheet of paper. They are house-hunting *Vibrio cholerae*—cholera bacteria—Kolter explains, and instead of poking their noses into closets or testing the faucets, they are probing the area with their flagella. "Individual cells will come and visit the surface for a while and then they leave. They haven't committed to staying. But eventually, they make a decision. When the flagellum feels that it's on that surface, it transduces a signal to the rest of the cell, and they begin to make the glue that sticks them." He taps the screen with a pencil, pointing to a bacterium that has come to a standstill. "So this guy *has* made the commitment." Seconds later he spots two more. "Or somebody around here, this one. And now you see that another one has joined it." Soon dozens have made similar decisions and settled down on the plastic surface.

This will be a family neighborhood now. It's time to trade in the compact car for a minivan or for bacteria settling onto a rock or a pipe, to replace flagella with a mode of transportation better suited to solid ground—a grasping finger of cytoplasm that microbiologists call a pilus. Here's how it works. Put your hand on a desk or tabletop and make a fist. Now, extend your index finger, bend your finger at the tip so you can grip the surface, and pull your hand up to your fingertip. Repeat—reach, grip, drag. Surface-dwelling bacteria use their pili like you're using your finger, to crawl, or "twitch," from place to place. And where might they be going? Why, to visit their new neighbors, of course.

On the screen, introductions are made by head-on collision, as two or three twitching bacteria slam into one another, run-ins that inaugurate enduring friendships. Traffic jams mature into cul-de-sacs. Chains of cells wind into tight spirals. The ties that bind these alliances are made of sugar, strands of sticky glycoproteins woven together to form a matrix that glues neighbor to neighbor and everyone to the surface. "They are like pioneers settling down to build a village," Kolter says. "The matrix keeps them from swimming away."

If a warm rock is a homestead, a warm rock dotted with such "microcolonies" is prime real estate. Families expand their territory, newcomers crowd in, late arrivals snatch the last parcels of open space. But there's more going on here than a population explosion. Putting down roots inspires a level of cooperation and civic responsibility unknown in fluid environments. Neighbors who were strangers only hours ago now collaborate to erect buildings, construct roads, distribute food, establish defenses, and organize trade. By the time the day is over, this village will mature into a bustling, well organized metropolis, a social arrangement that researchers such as Roberto Kolter call a "biofilm."

Step on a biofilm straddling a rock in a stream and all you'll notice is a slippery scum. Fly over the surface with Roberto Kolter at the helm of a confocal scanning microscope—an instrument that uses a laser beam to scan a specimen one focal plane at a time and then integrates the slices to reconstruct a three-dimensional image—and you'll get an aerial view of the skyline below the slime. Pillars and mushrooms jut up from the surface, high-rises that are home to millions of bacteria. A network of canals weaves around and between the pillars. These canals—in reality, fluid-filled channels dredged by growth-inhibiting secretions oozing from the bacteria themselves—form a primitive circulatory sytem, piping nutrients to those living in the high-rises and draining away dissolved waste products.

Cultural diversity and economic inequality, as characteristic of bacterial cities as our own, give rise to a patchwork of ethnic neighborhoods reflecting the uneven distribution of resources, regional differences in gene expression, and a willingness to accommodate multiple species. "In different areas of the biofilm, gene expression is different, so the bacteria take on different jobs. One guy is the cabinetmaker and one guy is the mechanic and one guy is the librarian," says Bonnie Bassler. Kolter concurs. "Chefs and grocers may settle together in the restaurant district, while musicians may settle near concert halls," he writes. For example, bacteria living near the surface of the biofilm, with easy access to oxygen and nutrients, are entrepreneurs, trend setters, lookouts. They grow and divide vigorously. Those with addresses in deeper layers, on the other hand, must make do with what little food and oxygen manage to reach the interior; the only job open to them is that of spore. But how do residents decide who is to explore and who is to sleep, where to erect a tower or dig a canal?

Silence is for pastoralists and loners. City dwellers need to talk. Bacterial cities print no newspapers and hold no public hearings, but chemical interchanges guide their behavior from the moment they begin to gather; as always, everyone's favorite topic of conversation is population density. Kolter explains: "The ones that stick signal to all of the others, 'We are at high density, so we are going to be more committed to staying here.'" Within the microcolonies, quorum-sensing signals then direct the construction of the biofilm's distinctive pillars and issue orders to excavate its interconnected canals.

Without signals, settlements cannot grow into species. Mutant bacteria, stripped of their ability to produce acyl-HSLs by genetic engineering, never progress beyond the microcolony stage. Instead of an orderly arrangement of towers and channels, these misfits, no longer capable of teamwork, pile on top of one another in an anarchic tangle devoid of architectural detail or social convention.

Because biofilms are rarely the work of a single species—the plaque that forms on the surface of teeth, for example, may be home to more than 500 varieties of microbes—city residents have to know generic terms as well as their own species-specific dialect. "Acyl-HSLs are involved, but other signals are also important," notes Kolter. Exchanges carried out in bacterial Esperanto—signals like Bassler's AI-2—help strangers who speak different languages among themselves collaborate to define neighborhoods, assign jobs, and recruit passersby.

City living has many advantages. Food can be harvested more efficiently, distributed collectively. Walls repel barbarian invaders. A solid foundation is an anchor in a storm. But one day the food supply may run out, the stream may be flooded with pollutants, the host will die. When living conditions take a turn for the worst, biofilm residents issue new signals, telling everyone that it's time to move on. Enzymes blowtorch holes in the web of glycoproteins, pillars and mushrooms collapse, channels dilate and then disappear, as clumps of bacteria pull free and swim away in search of greener pastures. Don't think of this breakdown as a disaster—deconstruction is part of the natural life cycle of bacterial cities. "If the bacteria were unable to escape the biofilm, the biofilm would, like an old apartment building, become a death trap," says Kolter.

Bacteria may live and work together as a biofilm as long as it's expedient; heart and lung bond for life. Still, this community of one-celled organisms shares many features with the complex tissues of plants and animals: a characteristic internal structure, a division of labor, a developmental program, all choreographed by chemical signals. Through the artful use of tribal dialects and a chemical lingua franca, bacteria can transcend their humble origins and mimic the great civilizations of larger organisms.

A PATTERN LANGUAGE

In his books, *The Timeless Way of Building* and *A Pattern Language*, architect Christopher Alexander describes a "vocabulary" of design elements he believes are essential to the construction of towns, neighborhoods, and buildings in harmony with their surroundings, because they are "deeply rooted in the nature of things." The words in this language are descriptions that Alexander calls *patterns*, configurations of architectural features that bear a particular relationship to one another:

> In a gothic cathedral, the nave is *flanked* by aisles which run parallel to it. The transept is at *right angles* to the nave and aisles; the ambulatory is *wrapped around* the outside of the apse; the columns are *vertical, on the line separating* the nave from the aisle, *spaced at equal intervals.* Each vault connects *four* columns, and has a characteristic shape, *cross-like* in plan, *concave* in space. The buttresses are run down the outside of the aisles, on the same lines as the columns, supporting the load from the vaults. The nave is always a *long thin rectangle*—its *ratio may vary between 1:3 and 1:6, but it is never 1:2 or 1:20.* The aisles are always *narrower* than the nave.

A pattern is not a blueprint, however; it not only describes what a structure should look like but also reflects the events and activities that typically occur there. For example, "each sidewalk . . . includes both the field of geometrical relationships which define its concrete geometry, and the field of human actions and events, which are associated with it." Because it unites form and function, a pattern, says Alexander, represents a solution to a problem that "occurs over and over again in the environment." And an ideal pattern is the *best* solution, one that describes "a *property* common to *all possible ways* of solving the stated problem . . . a deep and inescapable property of a well-formed environment."

In our spoken and written language, nouns, verbs, adjectives, and other parts of speech—solutions to the design problems associated with converting thought and experience into sounds: the naming of objects, the representation of tense, the subtle nuance of description—can be linked in ways specified by the rules of grammar to form larger patterns, or sentences. Similarly, elemental architectural patterns—entrances, hallways, windows—can be combined to generate larger structures that maintain sensibility and harmony, provided they follow a prescribed set of rules, "patterns which specify connections between patterns." A well-designed building, in other words, is like a grammatically correct sentence, an orderly arrangement of elements that "makes sense" because it follows the rules of this three-dimensional syntax.

In the turbulent incubator of a primeval world, the one-celled ancestors of all life forms conceived and nurtured ways to describe the physical world, as well as the social environment, in the language of chemistry. Chemical signals, and the machinery that cells evolved to detect and interpret them, became life's parts of speech; the conventions observed when these elements were grouped to form pathways, the rules of biological syntax; the sequences constructed according to these rules, grammatically correct statements capable of informing behavior. This language, based on molecules rather than sounds—particularly proteins, folded, coiled, and looped into "configurations of architectural features that bear a particular relationship to each other"—has a three-dimensional aspect missing from spoken language. It is, in a sense, a pattern language, featuring design elements crafted from carbon and hydrogen rather than wood, stone, or brick, which represent solutions to the problems encountered by cells as they evolved mechanisms to collect and transmit information about the outside world:

Problem: Finding the right words to start a conversation.

Solution: Make do with what you have. Cells didn't have to look any further than their own doorsteps to find molecules, like *S*-adenosylhomocysteine, that could be recast as signals, or methods that could be co-opted to make new ones, like the protein technology that Gram-positive bacteria used to craft oligopeptides.

Problem: Say what?

Solution: Get in shape. Protein topography solved the problem of discriminating signals, with unique combinations of ridges, clefts, and loops that could easily tell serine from aspartate, an autoinducer made by the same species from an autoinducer made by a stranger.

Problem: The cell membrane.

Solution: The transmembrane receptor. Receptor proteins, with a sensor component to recognize the signal, a lipid-loving segment to insinuate it in the membrane, and a cytoplasmic component to broadcast the news of signal binding inside the cell solved the problem of ferrying information across an impenetrable barrier.

Problem: Delivering the mail.

Solution: Teamwork. Receptors could have been linked directly to the intended recipient of the message—a gene, a rotor protein. But cells found a more elegant solution in the relay, exemplified by the Che protein sequence of *E. coli*. Extra signaling proteins increase complexity but more than make up for the imposition by providing opportunities to amplify and regulate the flow of information along the way. And by mixing and matching receptors and relays, as *E. coli* does with its four receptors and one set of Che proteins, a cell can economize, creating multiple signaling pathways with fewer proteins—a feature exploited to maximum advantage, as we will see, by the cells of higher animals.

Problem: Staying alert.

Solution: Redecorate. An informant who chatters relentlessly stops being informative—and so would a signaling mechanism that could be turned on but couldn't be turned down or turned off. A notch on a receptor or relay protein where a methyl group, a phosphate, or another protein could be attached allowed cells to fine-tune signal transmission. As a result, signaling pathways could remain responsive over a wider range of signal concentrations and react to changes in signal intensity rather than signal concentration.

Dogs and cats have devoted owners, horses and dolphins their aficionados. Bacteria have scientists like Bonnie Bassler and Roberto Kolter. To them bacteria are more than the occupants of the lowest rung on the ladder of life. "Bacteria are incredibly sophisticated organisms," insists Kolter. "They have been evolving much longer than us. I don't like the term 'primitive,' and I don't like to think of bacteria as fossils. If you think of diversity as the number of genes, bacteria as a group have a hundred times more than animals, spread all over the planet. In this sense they are much more complex."

Bassler agrees. "Everybody talks about them as being the simplest thing. I think they are the most sophisticated organisms. There's just no slop. They're perfected for every niche that they live in. They don't have a nucleus or organelles, and they still do everything we do."

Our bodies may be breathtaking in their sophistication, yet bacteria—life crammed into the smallest, most efficient package—have come up with the same answers to life's most challenging questions. They use the same genetic code, live by the same metabolic reactions. And like our own cells, they talk to one another, using a chemical pattern language governed by a common set of grammatical rules. These conversations, an extension of sensory mechanisms developed to coordinate responses to environmental stimuli, allow them to socialize as our cells do, to build communities with many similarities to complex differentiated tissues, and, in so doing, to enjoy improvements in the quality of life—better nutrition, more stability, safety from predators—that are the perquisites of multicellular living.

By exploiting the distinctive architecture of proteins and their affinity for a select group of signaling molecules, bacteria have crafted a language, talked their way to becoming the longest-running life forms on earth, and set the stage for more ambitious adventures in multicellularity. Words have changed and vocabularies grown but the basic syntax of the language of life—the signal-receptor-relay sequence illustrated in elegant simplicity by bacteria—has been conserved across kingdoms; it is also the foundation of the sophisticated biological sentences of higher organisms. A proven solution to a universal problem, this sequence and the elemental patterns that compose it truly describe "deep and inescapable properties" of communication in all cells.

BUILD IT AND THEY WILL TALK

Take a deep breath. Now thank the former bacteria living in each and every one of your cells that the oxygen you just inhaled sustains you rather than kills you.

The hardy cells that pioneered life were remarkably self-sufficient, capable of building their own walls and membranes, re-creating their own DNA, manufacturing their own proteins. But one necessity—the energy needed to power essential chemical reactions—could not be generated by even the most industrious and had to come from outside sources. Life's first cells made do with resources they found close at hand, tapping the chemical energy of inorganic compounds such as hydrogen sulfide—effective, but limited. Their progeny, however, discovered an inexhaustible alternative directly overhead: sunlight, earth's most abundant energy source. Chlorophyll, a green pigment able to convert light energy into chemical energy, allowed these innovators to draw on the sun's plentiful energy reserves to drive the synthesis of carbohydrates they could burn as fuel.

It was a clever and effective way to make a living, fashioning your own food from sunlight, water, and carbon dioxide. But photosynthesis also had a downside. The chain of chemical reactions that culminated in energy-rich sugar and starch also generated a noxious pollutant: oxygen, not only useless but dangerous to organisms that had evolved in its absence, wanton in its disregard for the integrity of biomolecules. Life on earth might have corroded like a rusting nail if not for evolution's uncanny ability to transform potential calamity into opportunity. A vanguard of enterprising organisms upgraded their metabolism with proteins that could harness the volatility of oxygen to increase the amount of energy extracted from carbohydrates. This invention—biologists call it "respiration"—not only protected vulnerable cellular constituents from oxygen toxicity but also enabled its inventors to realize a significant improvement in energy efficiency.

So some lived on geochemicals, some on sunlight. And others lived off the largess of others, solving their own energy problems by devouring their neighbors. Earth's first predators, they had no teeth or claws but used their own bodies to capture their prey, embracing, then engulfing and digesting it. A meal of oxygen-loving bacteria had additional benefits for one of these lucky predators, however. Just as "it was the horse's value as a mount, not as a meal, that quickly came to dominate its relationship with man," it was the victim's cache of respiratory enzymes, rather than its nutritive value, that proved more valuable to the predator. Instead of being dismembered and digested, these captives were permitted to survive and reproduce. Generation by generation, their descendants gradually ceded most of their DNA to the predator's genome; trading their freedom for a free ride, the aerobes became permanent houseguests, the cellular constituents we know today as mitochondria.

The organism that domesticated a microbe it should have digested was odd in other ways as well. While the sulfur eaters and photosynthesizers preferred an open floor plan and an informal lifestyle—their DNA rolled and stored in a corner, their proteins unbounded by walls or closets—this creature was a compulsive organizer. Its DNA was enclosed in a membrane-bound bubble, a chance involution of the cell membrane that gave it and its successors their family name: the *eukaryotes* (from the Greek *charyon*, meaning nucleus). An advance comparable to the invention of pottery, membranes grew so popular, in fact, that eukaryotes would ultimately appear to have splurged at an evolutionary Tupperware party, stretching, wrapping, and sealing a pliable membrane to form compartments of every shape and size. While they were adding interior walls, they stripped away the stiff outer cell wall favored by prokaryotes. In its place they erected an internal scaffolding of protein rods and filaments, a so-called cytoskeleton, that gave shape to their soft bodies but permitted them to move freely about.

A new social contract arrived along with the new furniture. Bacteria, eager to cohabitate, have never married—even those who settle in cities don't hesitate to move on if times get tough. But these eukaryotes, just as quick to recognize the many advantages of communal living, lived by the motto "until death do us part." Their alliances were indissoluble, their commitment to their colleagues, unequivocal. Some—these would become plants—decided good fences make good neighbors after all and surrounded each of their cells with a cocoon of cellulose. Others—the clan that biologists call "metazoans" and we know as animals—forswore rigid walls that prevented real intimacy for good. They dared not only to collaborate with their neighbors but also to touch and even embrace them.

A revamped metabolism, the freedom to bend and move, a lasting marriage, and a physical form that enabled physical contact were more than cosmetic changes—they were adaptations that supported a revolutionary new lifestyle. Yet they also led to formidable new problems in organization and management. For solutions to these problems, eukaryotes turned to a tried-and-true faculty: the gift of speech, embodied in the chemical language of receptors and kinases.

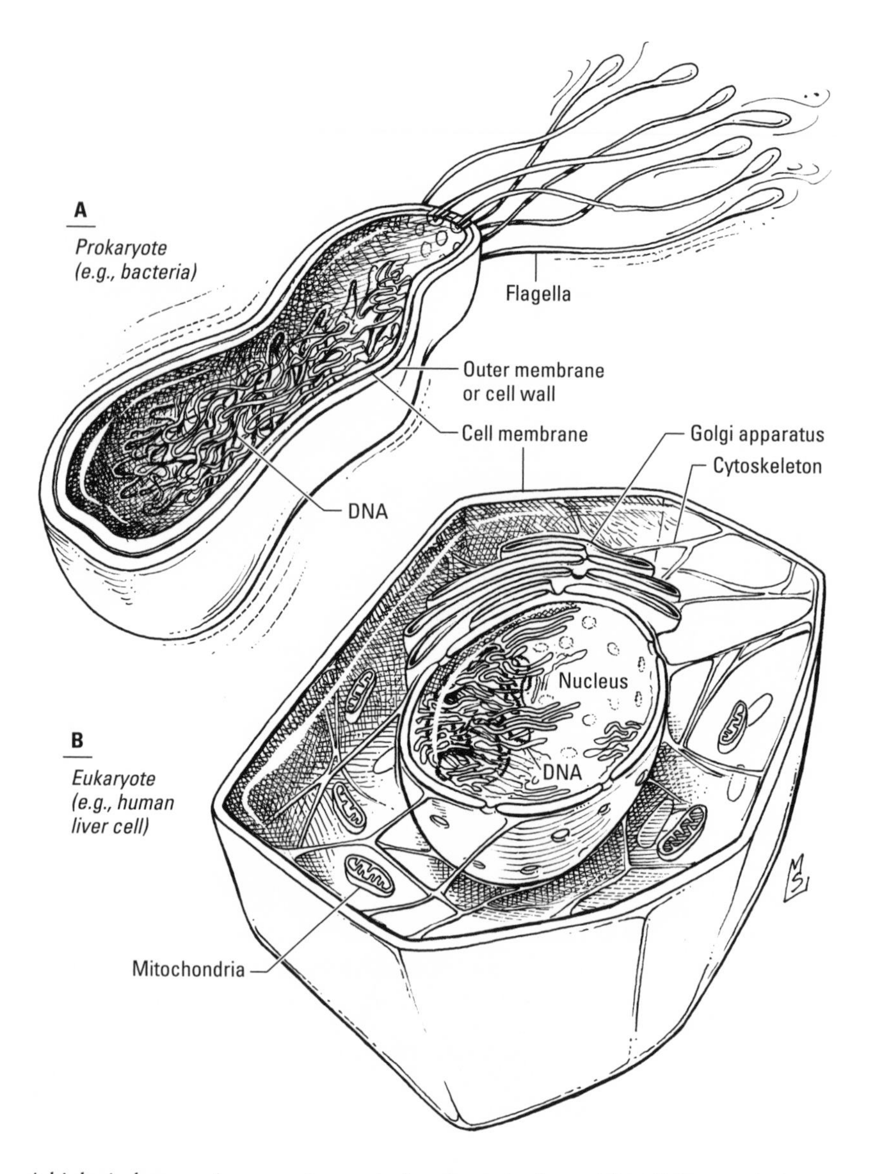

A biological generation gap. *A*, a typical prokaryotic (bacterial) cell. *B*, a typical eukaryotic cell, illustrating the nucleus, the internal system of membranes, and the Golgi apparatus, an example of the membrane-bound organelles.

FROM THE GROUND UP

I turn on my cell phone and instead of "Welcome," the screen announces "Holla!"

Ordered to erase the welcome message "can't knock the hustle," my older daughter Jennifer has substituted this. "'Holla'?" I ask. "Don't you mean 'Hello'?"

"No. It's 'Holla'."

"What's with that?"

"It's just something we say. Like, if somebody scores at a basketball game, everybody yells 'Holla!' Or if I see one of my friends on the stairs, we go 'Holla!'"

"So if I saw one of my friends in the mall, could I yell 'Holla, Marie'?"

Jenny rolls her eyes. "Only if you wanted to sound like an idiot."

"What if I saw you in the mall?"

"Don't go there."

"So when do I use it?"

"*You* don't. You're my mother, not my friend."

As any adolescent can tell you, parting ways with the adults isn't easy. You can't live somewhere else, you don't have a full-time job, and even if you can drive, it's their car. One way you can put distance between yourself and your parents, however, is to update their language. You can throw away some words, change the meaning of others, and add new ones to put your stamp on every conversation.

Eukaryotes—you could think of them as "Generation E"—had outgrown old-fashioned ideas like oxygen phobia and unprotected chromosomes, but they still had to share living space with bacteria, the older generation. Tinkering with language was one way to distance themselves from their elders, to establish their independence and keep the old folks from learning every detail of their personal lives—information that could be used against them. "As soon as eukaryotes evolved, they had to deal with bacteria," notes E. Peter

Greenberg, a microbiologist at the University of Iowa College of Medicine. "They needed signals bacteria didn't recognize, and that's probably why our signals are different from their signals."

But an identity crisis was only one of many factors inciting multicellular eukaryotes—particularly metazoans, or animals—to tamper with the language of life. First of all, they were growing faster than my daughters' wardrobes. In a world still dominated by single-celled organisms (both eukaryotic and prokaryotic), bigger was better; as biologist John Tyler Bonner writes, "There is always an open niche at the top of the size spectrum; it is the one realm that is ever available to escape competition." Selection for size, Bonner continues, goes hand in hand with "a selection for a better integration, a better coordination of the adhering cells." The small, limited to brief exchanges with their immediate neighbors, didn't need an expansive vocabulary or long-distance phone service. The cells constituting larger organisms, however, had to keep in touch with distant relations as well as those next door. Existing options just wouldn't do; they needed new kinds of signaling molecules, sturdy enough to survive a trip from one part of the body to the other.

For animal cells the freedom to embrace also established a new forum for discussion. Adrift in the primal ocean, they resisted dispersal by clinging tenaciously to each other in a continuous layer known as an epithelial sheet—life's first true tissue. In contrast to a biofilm, which conforms to the curvature of a rock, the edge of a tooth, or the surface of a pipe, an epithelial sheet could adopt any shape it chose, even curl up and anneal edge to edge to form a tube or sphere. Within such a hermetically sealed compartment, conditions could be specified by the organism, not left to the whims of a capricious environment, and cells could speak without their words being diluted and dissipated by every passing wave. Cellular behavior could be micromanaged by editing the content of this internal milieu—providing, of course, the organism had the words to spell out exactly what cells should do and when.

As if a population explosion and an inner life weren't enough to manage, there was the building program. Sheets could be fashioned into shapes, populations arranged to form patterns, and tissues assigned specialized tasks, but external features and internal organs did not form spontaneously. They had to be built from scratch, starting with a single fertilized egg—a feat that would have been impossible without contractors and career counselors. To shape head and limbs, gut and heart, brains and bones, multicellular animals needed road maps and directions, words of wisdom to keep embryonic cells on track as they searched for their destiny.

With their lockboxes for chromosomes, their bold exploitation of oxygen, their compulsive use of membranes, their internal skeletons, and their odd ideas about commitment and intimacy, animal cells were as different from their elders as low-rise-jeans-wearing, multiply pierced, skateboard-toting, electronically enhanced modern adolescents are from theirs. Yet the more things changed, the more they stayed the same. Cells, whether prokaryote or eukaryote, had to navigate the same capricious environment, face the same challenges collecting and disseminating information. The availability of resources and the presence of danger, the quality of life and the actions of neighbors, still had to be encoded in a form understandable to genes and proteins, still had to be transferred across an impermeable plasma membrane, still had to be delivered to targets that organized responses. So it's not surprising that while the conversations of eukaryotes sparkled with new ideas—solutions to the new problems created by their secession from the prokaryotic lineage, their quest to occupy the "niche at the top of the size spectrum," and the inherent difficulties associated with constructing and operating a larger, more complex body—much about these exchanges was strangely familiar.

But how could evolving eukaryotes be both traditionalists and trendsetters, as plain spoken as bacteria yet at the same time profoundly more articulate? The answer is simple: they were brilliant architects. Answers to age-old questions could still be found in a common pattern language, in archetypal patterns based on the architecture of molecules (particularly the ornate topography of proteins), in sentences that followed the same law-abiding internal organization as those of prokaryotes. Working from the ground up, cells extended and updated this library of fundamental design elements to adapt the chemical language of life to meet their growing needs. New signals, new substrates, new patterns, new combinations enhanced language skills, but the real differences between the language spoken by eukaryotic cells and the language of their prokaryotic counterparts lay in the scope not the structure of that language, in the breadth and elegance of the vocabulary, and the intricacy of sentences, rather than in the wholesale redesign of fundamental mechanisms.

Linguists debate the existence of a "universal grammar," an internal structure common to all languages, genetically programmed into the human brain. For cell biologists there is no debate. The language they study, the pattern language of cells, undeniably possesses a common internal structure. Elemental patterns and established rules of syntax, as familiar to *E. coli* as they are to human beings, constitute *the* universal grammar, part of the legacy of every cell.

A SURFEIT OF SIGNALS

Feeling under the weather? Tired, feverish, congested? Today, a doctor would probably attribute your symptoms to this season's flu virus. If you were a patient in ancient Greece, however, your physician would have known nothing of viruses and infections. He would have blamed the problem on an imbalance in bodily secretions called humors, each representing one of the four elements thought to compose all matter: yellow bile, associated with fire; phlegm, cold and wet like water; blood, corresponding to air; and black bile, dark and inert as earth.

The Age of Mythology gave way to the Age of Reason, and the

humors of the Greeks were relegated to museums along with the statuary and temple artifacts. But secretions responsible for health and well-being are still very much with us; today, we know them as "hormones," from the Greek *hormao*, meaning "I excite." Carried from speaker to target by the blood, these sturdy chemical messengers were the perfect solution to one of the quintessential problems facing multicellular eukaryotic organisms—coordinating the behavior of cells separated by long distances.

Bacteria may have been the first to use chemical signals, but the first chemical signals discovered by scientists were hormones, in particular the secretions of the so-called endocrine, or ductless, glands: the thyroid and the adrenals, the pancreas, the testes, and the ovaries. The knowledge that these glands had far-reaching effects on body functions predates even the Greeks, for farmers and herders have been familiar with the profound changes in appearance and behavior that accompany castration, the surgical removal of an endocrine gland, for millennia. And the idea that these effects were mediated by glandular secretions took shape more than 200 years ago, when Theophile de Bordeau, court physician to Louis XV, speculated that "each gland . . . is the workshop of a specific substance that passes into the blood, and upon whose secretions the physiological integration of the body as a whole depends." By the middle of the nineteenth century, scientists could cite clinical observations that supported Bordeau's hypothesis. "A diseased condition" of the adrenal glands, for example—now known as Addison's disease, after the English physician Thomas Addison, who published the first description of the condition in 1855—disrupted cardiovascular function, digestion, and appearance: "The leading and characteristic features of the morbid states to which I would direct attention are anaemia, general languor and debility, remarkable feebleness of the heart's action, irritability of the stomach, and a peculiar change of color of the skin . . . with the advance of the disease . . . the pulse becomes weaker and weaker, and without any special complaint of pain or uneasiness the patient at length gradually sinks and dies," while damage to the thyroid gland was toxic to the mind as well as the body—"cretinism" they called the result.

Cries of help from tissues deprived of some vital elixir only the adrenal gland or the thyroid gland could supply, some argued. But until someone actually isolated such a substance, any conversations initiated by the endocrine glands would remain confidential, and the simplest farmer would know as much about cellular communication as the most learned academic.

On an otherwise ordinary day in 1894, Edward Albert Schäfer, professor of physiology at University College London, was recording the blood pressure of a dog as part of a routine experiment when he was interrupted by a stranger with a vial of a mysterious substance and an urgent request. Dr. George Oliver introduced himself—taking care to note that he too was a former student of Schäfer's revered mentor, William Sharpey—and explained that when he was not treating patients he spent much of his time in a modest laboratory he'd set up in a back room of his home, designing medical instruments he then tested on family members. His latest invention was an "arteriometer," designed to measure the diameter of a blood vessel through the overlying skin; to try out this device, he had injected his son with glycerin extracts of various animal glands and recorded changes in the diameter of the boy's radial artery.

Oliver admitted that most of the extracts had done nothing. But one, prepared from the adrenal glands, had triggered a spectacular constriction of the artery. Didn't the professor agree that this might mean the extract contained one of the glandular secretions so keenly sought by scientists? Oliver handed the vial to Schäfer and exhorted him to inject some of the liquid it contained into the dog lying on the table. And Schäfer—perhaps merely curious, perhaps a bit intimidated by Oliver's fervor—agreed:

> So, Professor Schäfer makes the injection, expecting a triumphant demonstration of nothings, and finds himself, like some watcher of the skies when a new planet swims into "his ken" watching the mercury rise in the manometer with surprising rapidity and to an astounding height, until he wonders whether the float will be thrust right out of the peripheral limb. So the discovery was made of the extraordinary active principle of the suprarenal gland.

Chemists—John Jacob Abel and Jokichi Takamine—took over next. They plied the crude extract with acids and solvents and ammonia; by the turn of the century, they had succeeded in purifying the "active principle." Abel called the substance "epinephrine," and Takamine "adrenaline," after the gland that secreted it.

Cells could talk—and over what distances!—by exchanging chemical signals, molecules produced and excreted by cells of the endocrine glands and carried to cells in target tissues by the bloodstream. But hormones (following the purification of adrenaline, chemists also isolated cortisol, the missing factor in Addison's disease, from the adrenal glands, as well as thyroxine from the thyroid gland, estrogen and progesterone and testosterone from the gonads, insulin and secretin from the pancreas) represented only one of the many new types of words that animal cells added to their vocabularies. Conversations with the cells next door did not become obsolete just because they couldn't be heard on the other side of the body, and so short-range signals, as well as novel ways of deploying them, proliferated as well. Development, for example, relied heavily on signaling molecules that never strayed from the neighborhood where they were made and secreted. Neurons, the cells comprising the nervous system, pioneered a new type of interface for conducting one-on-one conversations, and recruited dozens of new substances—small molecules, peptides, even gases like nitric oxide—to encode their messages. Membrane-bound signals set up lines of communication between the gelatinous matrix surrounding cells and the cytoskeleton supporting them. Cells of the immune system improvised signals from the dismembered bodies of pathogens. Even death had its own dialect.

To find these new signals, eukaryotic cells, like bacteria, didn't have to look any farther than their own backyards. Built from chemicals, operated and maintained by chemicals, and fueled by chemical reactions, they lived in a dictionary. If a stray metabolite, a common nutrient, or a by-product already on hand wasn't quite right, or if they were still at a loss for words, they could always turn to proteins for ideas. Thanks to a constantly evolving genome, new models were always coming on the market—enzymes born to be wordsmiths or precursors pregnant with prospective signaling peptides. Adrenaline, for example, is just the amino acid tyrosine dressed up in fancy clothes, the neuronal messengers norepinephrine and dopamine, byproducts of the same synthetic process. Testosterone, the hormone that creates boys and turns them into men, begins life as cholesterol, as do estrogen, progesterone, and cortisol. Many growth factors are proteins, insulin and secretin are peptides—new signals but manufactured according to time-honored procedures.

Biodiversity—the mosaic of life forms populating an ecosystem—confers the resilience that enables that system to respond to a variety of environmental challenges. A diversity of signaling molecules provided evolving eukaryotic organisms a similar flexibility to meet the challenges posed by new internal contingencies. Whether the situation demanded a subtle phrase appreciated only by intimates, a hearty greeting to faraway relations, words with a lasting impact, or words quickly forgotten, their cells found a way to say it—an innovation essential to the survival of the vulnerable ecosystems contained in their bodies.

COULD YOU REPEAT THAT PLEASE?

Paul Ehrlich's nineteenth-century professors called him disorganized, incorrigible, unteachable. A modern educator would probably call him a "visual learner." While other medical students spent long hours in the library preparing for exams, Ehrlich spent hours in the laboratory coloring tissue sections with dyes created by the burgeoning German chemical industry, far more fascinated by the unique patterns they created—one stained only the cytoplasm of cells, another only nuclei, a third bacteria, the one in the bottle next to it, the cells they attacked—than by the anatomy of the hand or the names of bones. Who would have thought such an indifferent student would ever become a doctor or that the diphtheria and tetanus antitoxins he began investigating in his own laboratory would interact selectively with the noxious secretions of these pathogens alone, a specificity that recalled that of the dyes he'd tinkered with in medical school?

Other professors called John Newport Langley brilliant. A contemporary of Ehrlich, Langley was a respected Cambridge physiologist, the first to describe the so-called autonomic nervous system that relays nerve signals to body organs, as well as a fellow of the Royal Society, president of the Neurological Society of Great Britain, editor of the *Journal of Physiology*, and the recipient of dozens of honorary awards. Who would have thought such a virtuoso would have anything in common with a man who was the despair of his teachers? Yet Langley was also fascinated by the selectivity of chemicals (drugs in this case), how each seemed to home in on some tissues and ignore others. Both nicotine and the South American arrow poison, curare, for example, were attracted to muscle—yet the first stimulated contractions, while the other blocked them.

Despite their differences, both men also came up with the same explanation for such specificity: physical attraction. A dye stained certain structures and not others, Ehrlich suggested, because it bound to some component found only in those structures. Diphtheria antiserum neutralized the poisonous secretions of diphtheria bacteria because it recognized a structural feature peculiar to that toxin; in the body, cells fell victim to bacterial assault because they also contained a feature—Ehrlich called it a "side chain"—that matched a loop or groove in the toxin molecule. Similarly, Langley proposed that drugs formed "compounds" with a "receptive substance" on the surface of the muscle cell and suggested that compounds like nicotine and curare canceled each other's actions because both vied for this same target.

"*Corpora non agunt nisi fixata*," Ehrlich concluded. "Agents cannot act unless they are bound." Pharmacologists agreed, and endocrinologists reasoned that hormones, too, probably exerted their biological effects by binding to receptors. The idea was so commonsensical, yet for half a century after Ehrlich and Langley first posited the existence of "side chains" and "receptive substances," receptors themselves remained as elusive as hormones once had. "Receptors had no basis in fact; they were more of a metaphysical entity. No one had any thought of what they might be," explains Nobel Prize–winning pharmacologist Alfred G. Gilman. Scientists could see their handiwork in muscle contractions or the constriction of an artery, but the tales told by such reactions were only hearsay, events several steps removed from the actual binding site. As a result, says Gilman, "People were spectacularly clueless about how signaling might occur."

Once again, adrenaline would animate a theory, replace metaphysics with molecules. Matriarch of hormones, it still had stories to tell, this time about the nature and behavior of receptors, beginning with the mysterious mechanism that translated the binding of a hormone at the surface of a cell into a response on the other side of the plasma membrane.

A body pumped up on adrenaline can't run on thin air. Fighting and fleeing are high-energy activities, fueled by the breakdown of glucose. Running out during an emergency could be catastrophic; fortunately, the liver maintains a reserve supply, harvested from the bloodstream in times of plenty and stored in polymer form as glycogen as well as an enzyme, glycogen phosphorylase, able to deconstruct the polymer. Congenitally lazy, however, glycogen phosphorylase spends most of its time sleeping, the sweet taste of sugar no more than a dream. Snug under a soundproof quilt of cytoplasm and membrane, the enzyme might never hear the clanging of a catecholamine if not for the efforts of the adrenergic (adrenaline-specific) receptor.

The receptor, Earl Sutherland learned, is a typical manager—it doesn't rush off and wake up glycogen phosphorylase itself; instead, it delegates the job. An apprentice at the time to biochemists Carl and Gerty Cori, Sutherland, along with fellow future Nobel Prize winners Edmond Fischer and Edwin Krebs, learned in the early 1950s that the adrenergic receptor activates a kinase (now known, aptly enough, as "phosphorylase kinase") and the kinase is the alarm clock that actually rouses glycogen phosphorylase. But if he wanted to know more about this relay, Sutherland realized, he was going to have to pull back the covers and find out how receptor and kinase interacted on the underside of the plasma membrane, after the interaction of receptor and adrenaline on the outside.

Disrupt the membrane and expect a receptor to behave normally? Biologists agreed one might as well crack open an egg expecting a live chick to hop out. "It was almost the definition of a hormone, that its effects could only be observed in the intact cell—if you broke up the cell, the hormone effect went away," observes Gilman. Fortunately, he adds, Earl Sutherland was a lucky man. Along with colleague Theodore Rall, Sutherland discovered that liver cells could be ground to a pulp without compromising adrenaline's ability to activate glycogen phosphorylase. In fact, Sutherland and Rall found that they could spin the slurry in a centrifuge to separate receptor and enzyme entirely (glycogen phosphorylase, a cytoplasmic protein, floated in the fluid portion at the top, while the membrane fragments containing the receptor settled to the bottom), put them back together again in a test tube, and recover a response to adrenaline as robust as that found in a living cell.

If the receptor was a protein—and Sutherland and Rall assumed it was—it ought to fall apart when heated. So as a control they added adrenaline, then boiled the membrane fragments to destroy the receptor before recombining them with the cytoplasmic fraction. But to their surprise the hormone had still managed to activate glycogen phosphorylase. They concluded that the sequence from receptor to enzyme must contain yet another go-between, and it was not a protein. This hardy molecule, Sutherland learned, was a nucleotide—an amalgamation of a sugar, a nitrogen-spiked moiety or "base,"and phosphoric acid—related to the ATP (adenosine triphosphate) cells relied on to power chemical reactions but bent into a ring and decorated with a single phosphate group instead of three, a structure reflected in its chemical name: "cyclic-3′, 5′-adenosine monophosphate," or "cyclic AMP" for short. He called this adrenaline-by-proxy a "second messenger" because it repeated the command first issued by adrenaline. "It was a biochemical, rather than physiological, demonstration of receptor activity, one of the first times you could think of receptors in real biochemical terms," concludes Gilman.

Within a few years it would be possible to think of receptors in even more intimate terms. The study of cell signaling was about to enter the Nuclear Age.

Hydrogen: small, light, invisible. Iodine: a reminder that life began in the sea. Carbon: black scaffold of life. Phosphorus: so volatile it ignites spontaneously. Sulfur: the biblical brimstone. Nuclear chemists can induce all of them to form radioactive isotopes, a transformation that changes their destiny—they will now inevitably, inexorably, disintegrate—but not their basic chemical nature. Recruited as substitutes for their naturally occurring, stable counterparts, radioactive atoms readily assume the same roles in chemical reactions. But they are as conspicuous in the finished product as the nonconformist neighbor who lets his lawn "go natural" and paints his shutters pink, meaning that the activities of the molecules harboring them can be monitored as readily as the movements of a wild animal fitted with a radio transmitter.

Radioactive tracers took the secrecy out of biomolecular behavior. Nutrients spiked with radioisotopes exposed synthetic pathways to public inspection. Radiolabeled amino acids recounted the life stories of proteins. Woven into DNA and RNA, radioactive nucleotides charted the course of cell division and the transcription of genes into RNA recipes for proteins. And with radiolabeled hormones and drugs acting as reporters, scientists could finally relish eyewitness accounts of the long-clandestine interactions between receptors and signaling molecules.

Cyclic AMP turned out to be a popular word. In the decade following Sutherland's discovery of the molecule, scientists discovered at least a dozen other hormones and neurotransmitters, in addition to adrenaline, that utilized it as a second messenger. In fact, by 1969 so many scientists had become interested in such receptors that it seemed as if "half the world at the time was studying cyclic AMP," recalls biochemist Robert Lefkowitz. The investigators supervising Lefkowitz's postdoctoral research at the time, biochemists Ira Pastan and Robert Roth, were part of the wave. They were especially intrigued by the idea of hunting down cyclic AMP–dependent receptors with the new radioactive tracers, and so "the project they cooked up for me was studying adrenocorticotropic hormone—ACTH—using ^{125}I-ACTH,"* he explains. The assignment led to a paper in the prestigious journal *Proceedings of the National Academy of Sciences,* "the first description of a receptor that acted through cyclase"—and the first entry in a scientific bibliography that now includes more than 500 publications.

But Lefkowitz had no intention of devoting his entire career to

*That is, the hormone embellished with a radioisotope of iodine. ACTH is one of several hormones secreted by the pituitary gland, located at the base of the brain.

ACTH receptors. His real passion was adrenaline, his goal to use similar "radioligand" binding assays to study its receptors. "On the emotional side," he notes, "I was a fellow in cardiology at the time and I wanted to do something which would have immediate relevance to clinical medicine." And on the practical side, peptide hormones "were harder to work with, and there were few compounds that interacted with their receptors. In the case of epinephrine, there were hundreds of compounds."

That "side chain" idea had filled Paul Ehrlich's head with dreams of miraculous cures—and the shelves of pharmacies with new medicines. Extrapolating from the specificity of his dyes and antitoxins, Ehrlich argued that it ought to be possible to identify chemical agents with a similar specificity for pathogens, "substances which have an affinity to the cells of the parasites and a power of killing them greater than the damage such substances cause to the organism itself, so that the destruction of the parasites will be possible without seriously hurting the organism." He called the concept "chemotherapy" and demonstrated its legitimacy with his discovery of the arsenic derivative arsphenamine, or Salvarsan, the first man-made anti-infective agent. Following the same line of reasoning, physiologists-turned-drug-prospectors panned for drugs that would bind to the "side chains" recognized by endogenous signaling molecules, replicating or blocking their action. In the case of adrenaline, the hunt had been especially productive, providing physicians, as Lefkowitz points out, with an entire pharmacopoeia of drugs attracted to adrenergic receptors: drugs to shore up failing hearts, reduce soaring blood pressures, cheer depressed brains, neutralize life-threatening allergic reactions, and dilate airways throttled by asthma.

Lefkowitz selected promising members of this library and tagged them with the radioisotopes ^{125}I or ^{3}H, then offered them to the red blood cells of frogs and turkeys, fat cells and lymphoma cells, membranes from heart and brain. He counted receptors and compared the rate and affinity of binding to the efficacy of the drugs in physiological assays. He set up contests between his radioligands and unlabeled adrenergic drugs, calculated their affinities (the strength of the interaction between drug and receptor), and used these values to construct pharmacological profiles of the receptors in various tissues. His profiles corroborated an odd feature of adrenergic receptors first described by pharmacologist Raymond Ahlquist in the 1940s: they had multiple personalities. Those located on blood vessels, for example, responded vigorously to adrenaline but only tepidly to the adrenaline-like drug isoproterenol (Ahlquist had called these "α [alpha] receptors"), while those in the heart were their mirror image, preferring isoproterenol to either epinephrine or its precursor norepinephrine (these he called "β [beta] receptors"). The binding assays differentiated α and β receptors more emphatically than bioassays; in fact, drug profiling suggested they could be subdivided even further: α receptors into two groups (α_1 and α_2); β receptors into three (β_1, β_2, and β_3).

Once researchers like Lefkowitz could get close to adrenergic receptors with binding assays, it also became realistic to think about extracting the receptor protein from the membrane and purifying it—using radioligands to track the protein during the rough-and-tumble purification process—dissecting it into its constituent amino acids, even cloning its gene. Not that receptors submitted to captivity meekly. Like other membrane proteins, they were maddeningly agoraphobic, loathe to abandon their familiar lipid-rich environment, quick to unravel and collapse if treated too harshly. Nonetheless, by 1982, Bob Lefkowitz and his co-workers had corralled the β_2-adrenergic receptor; by the end of the decade, they had purified β_1- and α-adrenergic receptors as well—proof that receptor subtypes were actually distinct proteins, not experimental artifacts.

Once they'd cloned the gene for the β_2-receptor and deduced something of its internal structure, it was not difficult to see why the protein had been so recalcitrant either—it was woven into the very fabric of the plasma membrane. Seven hydrophobic segments, each coiled into a helix, pierced the membrane, connected by loops of hydrophilic residues that protruded into the watery environment on either side. Viewed in cross section, the receptor appeared to have been sewn in place by an amateur who forgot to secure the ends of the thread, a baggy running stitch of a protein with one end draped across the surface of the membrane and one dangling in the cytoplasm.

Earl Sutherland discovered adrenaline's intracellular surrogate. Bob Lefkowitz put its receptors on display. Martin Rodbell added one receptor and one enzyme and came up with three components.

A COMMUNITY OF EFFORT

The son of a Baltimore grocer, Rodbell might have been a chemist—he did his doctoral research on phospholipids, the most important ingredient in your standard cell membrane. Or he might have been a poet—he'd studied French existential literature in college and liked to embellish scientific presentations with original poems. He might even have become a mathematician or an engineer, given his fascination with the avant-garde field of cybernetics—the term invented by Norbert Wiener to describe mechanisms common to machines and biological organisms—particularly the branch aficionados were calling "information theory." Instead, he chose endocrinology. He would study the compositions of cells rather than those of poets, by translating the language of life into the language of machines.

Where some saw a liver cell, Rodbell saw a tiny computer. "The living cell," he would later write, "is in essence a communication device, built primarily of organic matter rather than the silicon of today's computers." Drawing an analogy to the circuitry common to modern electronic devices, Rodbell proposed that "signal transduction"—the transfer of information from the external environment to the inside of the cell—required a sensor component, or "discriminator," to detect chemical signals, the data in this system; an output component, or "amplifier"; and, in addition, a third component, a "transducer," that did the actual work of converting the signal detected by the discriminator into a form that could be understood and converted into a response by the amplifier. In the adrenaline-adrenergic receptor-cyclic AMP system, for example, the receptor played the role of discriminator, and the enzyme that generated cyclic AMP, which had been christened "adenylyl cyclase," was the amplifier. What Rodbell's transducer might be—if it even existed in living cells—was anyone's guess.

The search for the transducer began with the receptor. Using the peptide hormone glucagon to stimulate adenylyl cyclase in liver cells, Rodbell decided to warm up with a few standard-issue experiments—radioligand binding studies with radioactive glucagon, assays to measure the output of cyclic AMP at different hormone concentrations—that should have been a mere formality. They weren't. Radiolabeled glucagon behaved strangely in the test tube, and the results of the binding studies did not mesh with the results of the cyclic AMP assays, even though both were carried out under the same conditions.

For Martin Rodbell, this apparent setback was a breakthrough. As he looked back over the experimental protocols, he noted that both assays contained ATP—the starting material for cyclic AMP and also the source of many a headache for biochemists. "Realizing from painful experience as a graduate student that commercial preparations of ATP contain a variety of contaminating nucleotides," he wrote afterward, "I tested many [other] types of . . . nucleotides." As he suspected, the "ATP" he'd bought from a scientific supply company was, in fact, contaminated with significant quantities of "another type of nucleotide," called guanosine triphosphate, or GTP. GTP would have been insulted by the suggestion it was a mere impurity, however. Adenylyl cyclase, it turned out, liked having its ATP spiked with GTP; offered the real thing, it quit making cyclic AMP.

Rodbell proposed that GTP was needed to jump-start the cyclase because this nucleotide manipulated the transducer to promote the orderly transfer of information from receptor to enzyme. What's more, using a radiolabeled analog of GTP, he identified a guanine nucleotide binding site in liver cells distinct from the binding site for glucagon. Subsequent studies showed that the transducer/binding site not only attached itself to GTP but devoured it as well, using a molecule of water like a crowbar to pry off a phosphate group and turn GTP into guanosine diphosphate, GDP. In other words, this component—Rodbell called it an "N protein"; his successors preferred the term "G (for guanine nucleotide) protein"—like protein kinases, was a phosphate-driven switch. The receptor was the finger that toggled the switch, adenylyl cyclase, the object of its machinations. In the absence of hormone, GDP was snuggled in the binding site, and the G protein was switched off. When the receptor got fired up by the binding of hormone, it drove the GDP out, allowing a molecule of GTP to take its place. Now the switch was on. The activated G protein stimulated adenylyl cyclase, then snapped a phosphate from GTP to regenerate GDP and reset the switch.

Even bacteria value such "GTPases"—who wouldn't find such a simple switch useful around the house? But only eukaryotes had the genius to incorporate them into signaling pathways. The transducer Rodbell discovered interceding for glucagon, subsequently christened "G_s" because it stimulates adenylyl cyclase, was, in fact, only the first of what has proven to be an extended family of such G proteins. Another, G_i, inhibits the production of cyclic AMP. A third, G_q, activates the enzyme phospholipase C, prompting the synthesis of another second messenger. The one known as G_t serves rhodopsin, the "light receptor" of the visual system, while others work hand in hand with receptors for hundreds of odors.

Rodbell referred to the transducer as a protein, but it was not at all clear that this element was, in fact, a separate molecular entity. "There was no compelling reason to believe that the guanine nucleotide binding site was on a separate protein; it could have been intrinsic to the receptor or the effector protein [adenylyl cyclase]," Alfred Gilman explains.

Gilman took up where Rodbell left off, recognizing that the only way to determine if the G protein was part of the receptor, part of the enzyme, or a third protein was to try to isolate it. "Al Gilman and I were on a parallel course for nearly a decade," says Robert Lefkowitz. "What I was trying to do with β receptors, he was trying to do with G proteins."

You could say Alfred Goodman Gilman was born to be a pharmacologist. His father, Alfred Gilman, was not only a respected professor of pharmacology—the first chair, in fact, of the pharmacology department at Albert Einstein College of Medicine—but also (along with Louis S. Goodman) co-editor-in-chief of one of medicine's best-known and most-respected texts, *The Pharmacological Basis of Therapeutics*, now in its 10th edition. "As my friend Michael Brown once said, I am probably the only person who was ever named after a textbook," Gilman the Younger notes.

As a child, Gilman, inspired by trips to the Hayden Planetarium, wanted to be an astronomer, but fate seemed to have hitched his star to adenylyl cyclase. Earl Sutherland (of cyclic AMP fame) was a family friend. He persuaded Gilman to accept a place in a new M.D.-Ph.D. training program at Western (now Case Western) Reserve University; there, Theodore Rall, Sutherland's former collaborator, became Gilman's graduate advisor. After receiving his degree, Gilman tried to switch to brain research, signing on as a postdoctoral fellow with neurobiologist Marshall Nirenberg at the National Institutes of Health. Instead, his new mentor put him to work designing a better assay for cyclic AMP. Gilman finally capitulated. He gave up on the brain and took on the task of purifying G_s instead.

Where Lefkowitz used radiolabeled drugs, Gilman used lymphoma cells with a curious genetic defect. The parent strain, a common laboratory cell line known as S49, suffered from what toxicologists would call an "idiosyncratic drug reaction"—exposed to the rush of cyclic AMP generated by adrenergic drugs like isoproterenol, they died in their dishes. The mutant variety of S49 cell survived. Their resistance to the lethal effects of cyclic AMP, Gilman and others learned, wasn't the result of a defect in adenylyl cyclase—the enzyme was fine. And the mutants had plenty of fully functional adrenergic receptors. The problem was at the level of the G protein; as a consequence, Gilman and colleague Elliot Ross reasoned, they should be able to determine if a protein isolated from normal cell membranes was, in fact, the sought-after G protein by introducing it into mutant S49 cells, treating the cultures with isoproterenol to increase cyclic AMP, and seeing if the cells lived or died. "What was needed was to rejoin the missing link," he explains.

By 1980, Gilman and his colleagues had succeeded in purifying a protein that met the criteria for a bona fide G protein—actually a protein alliance, for Gilman discovered that a G protein is really a confederacy of three polypeptides. Gα, the largest, is also the busiest: the subunit that initiates the collaboration with receptors, it also binds guanine nucleotides, strips GTP of its phosphate, and activates adenylyl cyclase (and other effectors). Gβ and Gγ are Siamese twins. In a dormant G protein the three subunits cling to each other, forming a single complex. When Gα ejects GDP, however, Gβ and Gγ get booted out as well. After Gα has gotten enzyme prodding and nucleotide trading out of its system, the others return to the fold, regenerating the original tripartite protein.

Once thought to be superfluous or at best, supportive, today the Gβγ duo is recognized to be an important intermediary in its own right, linking G protein–coupled receptors to other signaling relays, in addition to the one headed by adenylyl cyclase. "It's now appreciated that the separation of Gα and Gβγ represents a real branch point in signaling," Gilman explains. "βγ complexes regulate downstream effectors as much as Gα does." When the G protein is inactive and all three subunits are bound together, he continues, "the interaction between α and βγ mutually occludes the surfaces that each uses for downstream interactions. Subunit dissociation liberates sites of interaction on both Gα and Gβγ."

In addition to the G protein amino acid sequence, Gilman and others now have pictures of one of these proteins, the inhibitory G protein G_i. Because G_i is among the rare proteins willing to strike a pose for crystallographers (most balk at the thought of sitting still in crystal formation), they have been able to reconstruct its three-dimensional structure from X-ray diffraction patterns and magnetic resonance spectra. They can show you Gα hanging from the plasma membrane by its fingernails and trawling for hormone-activated receptors. Look at Gβ and its unshakeable sidekick, Gγ, straddling Gα's switch region; no wonder they're popped off when that switch snaps shut. Fortunately, as you can see, Gβ, shaped like a seven-bladed propeller, looks as if it's ready to travel. And Gγ is a born passenger. Slung across its partner's shoulder, it just holds on and takes in the scenery, one end trailing like a ribbon in the breeze.

"Put it down in my mind to great insight and intuition that a third piece had to be there," says Gilman, who shared the 1994 Nobel Prize in Physiology or Medicine with Martin Rodbell for their work on G proteins. Rodbell himself credited teamwork, comparing his discussions with co-workers to the conversations between cells. "Biological communication," he said in his acceptance speech, "consists of a complex meshwork of structures in which G proteins, surface receptors, the extracellular matrix, and the vast cytoskeletal network within cells are joined in a community of effort, for which my life and those of my colleagues is a metaphor."

GPCR. No, it's not the name of a new alternative rock band. It's an abbreviation for "G protein–coupled receptor," one of the names researchers have given the β-adrenergic receptor, the glucagon receptor, the ACTH receptor, and all the others that communicate through a G protein transducer. You could also call them "7TMRs," for "seven transmembrane receptors"; if you don't like abbreviations, call them "heptahelical receptors," or "serpentine receptors," for the way they flow sinuously across the plasma membrane. You'll never be wrong, for all 800-odd proteins that belong to the G protein–

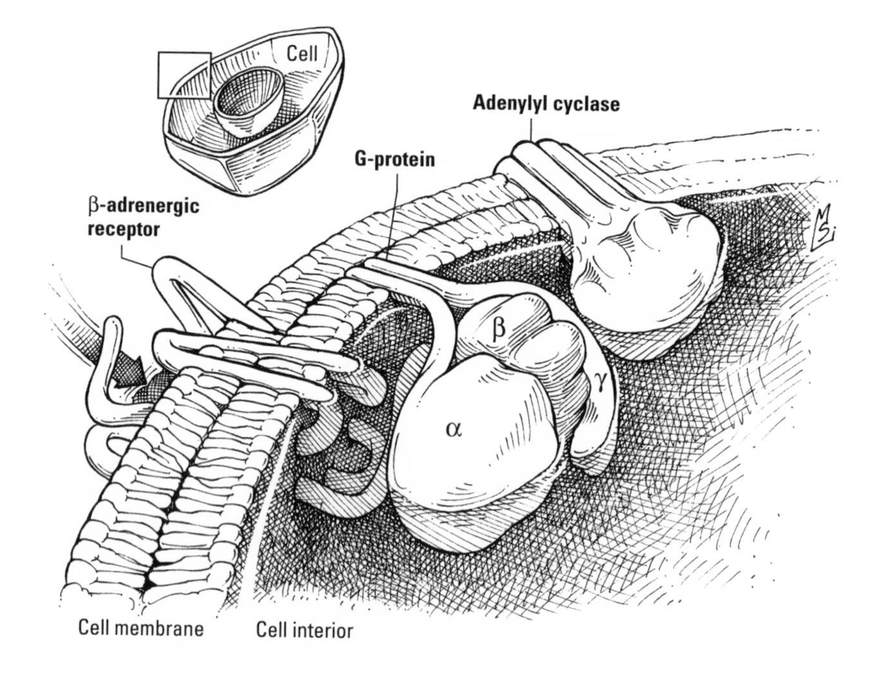

The β-adrenergic receptor, a G protein-coupled receptor. Like other members of this expansive receptor family, the β-adrenergic receptor is actually the first element of a three-protein phrase that also includes a G protein (in this case, G_s), a GTP-driven switch, and an effector, here the cyclic-AMP-generating enzyme adenylyl cyclase. The binding of signal to the receptor triggers the release of GDP from the Gα subunit of G_s, subsequently replaced by GTP, and the dissociation of Gα from the Gβ and Gγ subunits. GTP-activated Gα then stimulates adenylyl cyclase, increasing the generation of cyclic AMP, while the Gβγ duo goes its own way and speaks to other signaling relays.

coupled receptor family have the same seven-helix, membrane-spanning structure. "We were shocked," insists Bob Lefkowitz, who discovered the pattern when he found that his newly sequenced β_2-adrenergic receptor and the G-protein-coupled light receptor, rhodopsin, were look-alikes. "It came as a complete surprise. But the importance was recognized immediately."

Animal cells haven't discarded sensor kinases or receptors that double as transcription factors. Nor did their innovations in the receptor department begin and end with G protein-coupled receptors; all told, scientists have described at least 16 different varieties of receptor listening for signals in these cells. But multicellular animals begin more sentences with a receptor-G protein pair than with any other introductory word or phrase. A new take on an old idea—the transmembrane receptor—GPCRs illustrate how multicellular eukaryotes used simple design elements as the basis for larger patterns that expanded their language skills. Versatile and productive, these receptor phrases can be found controlling every type of physiological function in metazoan organisms, from simple housekeeping chores like the disposition of glucose to the intricate machinations of the brain.

SNAP TO IT

If cells were editors, they'd point out that the complete sentence, as spoken by the adrenal gland to the liver cell, goes like this: "adrenaline—β-adrenergic receptor—G_s—adenylyl cyclase—cAMP—cyclic AMP protein kinase (or "PKA"; heading the relay between receptor and target, it's the first protein to heed cyclic AMP's call)—phosphorylase kinase (Sutherland's kinase, it's phosphorylated and activated by PKA)— glycogen phosphorylase—THEN—Gβγ—G protein-coupled receptor kinase ("GRK," for short)—β-adrenergic receptor (phosphorylated by GRK)—β-arrestin—OR—PKA— β-adrenergic receptor (phosphorylated by PKA)—G_i—AND THEN"—well, more about that in a moment.

It's a mouthful, compared to the terse commands issued by prokaryotes. Yet despite the extra words, the speech still observes the same conventions as *V. fischeri*'s demand for light or the spore-making commands of *B. subtilis*. Hormones, G proteins, second messengers—these are new and creative, but beginning a sentence with the binding of a signal to a receptor is a tradition observed anytime, anywhere, that cells gather together. Kinases remain the backbone of the workforce; they have simply chosen new substrates; specifically, serine, threonine, and tyrosine. Messages are still handed from protein to protein, only the number of steps from receptor to target has increased.

Admittedly, creativity and eloquence posed challenges. Composed haphazardly, with proteins out of sequence, longer sentences could easily degenerate into nonsense. The mail could languish undelivered if the elements of a relay were scattered to the four corners of the cell or if an essential kinase was trapped in the crosstown traffic of a busy cell. And signaling proteins used once and then discarded were a senseless luxury; cells would have needed even more proteins—and more genes—every time they had something new to say.

Coherence, order, and efficiency demanded a way to connect and share signaling proteins. The answer? Conjunctions. Taking advantage once again of the unique structural features of proteins, eukaryotic cells came up with a design element that was the molecular equivalent of "and" or "or." A pattern without precedent, this new kind of signaling protein neither manipulated phosphates nor activated enzymes but evolved specifically to forge connections. Variously called linker, adaptor, or scaffolding proteins, the new connectors were mortar and magnet to eukaryotic cells, bridging the gaps between signaling elements, preventing errors that could lead to catastrophic misunderstandings, and conserving genetic resources by increasing the versatility of the proteins they connected.

Imagine you've just finished painting the living room, and now it's time to fold up the drop cloth you put down to protect the carpet. You could wrestle with the damn thing yourself. But wouldn't the job be easier if you had a helper, folding from one end while you folded from the other?

If proteins were painters, they'd have that floor clear in no time—for them, folding is always a group effort. That's because the amino acids that make up a protein are organized into blocks called domains. Each domain, a segment of 40 to 350 amino acids, assumes responsibility for folding that part of the protein, then works with the others to complete the job.

Domains, biologists believe, may have started off as proteins themselves. Over time the genes encoding them were inadvertently duplicated or fused with other genes to form a composite gene that encoded a new, larger protein. Although they had lost their independence, domains integrated into bigger proteins still folded themselves into the same familiar shapes and continued to carry out the functions they'd perfected while living on their own. So-called *interaction domains*, for example, retained the inherent ability to recognize and dock with a "consensus sequence" on another protein. As a consequence, a protein that acquired an interaction domain during the course of evolution inadvertently acquired a new associate as well. Such genetically engineered congeniality was not without risks—there was no law to prevent the same interaction domain from being spliced into several proteins. Conflicts would have been inevitable if not for the fact that in each protein the domain also recognized a unique configuration of amino acids flanking the consensus sequence, a second binding site responsible for choosing the one best match from the pool of potential partners. But even if promiscuity was forbidden, a protein could practice polygamy by adding more interaction domains, allowing it to bond simultaneously with several well-chosen consorts.

As the population of proteins needed to operate increasingly complex organisms grew, interaction domains emerged as an important locus of social control. Partnerships based on these domains organized proteins into teams and networks, directed them to specific locations within the cell and regulated their activity. Portable and self-sufficient, domains facilitated experimentation with new combinations of proteins and graced old warhorses with new flexibility. Yet at the same time they minimized demands on the evolving genome; as molecular biologists Mark Ptashne and Alexander

Gann argue in their book, *Genes & Signals*, "to evolve increasingly complex biological systems, it may not be necessary to invent many kinds of new gene products. Rather, more sophisticated functions can be achieved, for example, by increasing the number of interactions that any one protein can make, through the reiterated use of simple binding domains, thereby expanding the possibilities for combinatorial association."

In organisms that needed to direct the behavior of a multitude of cells, interaction domains that "expanded the possibilities for combinatorial association" were the perfect way to expand signaling pathways.

Molecular biologist Tony Pawson, of the Samuel Lunenfeld Research Institute at Mount Sinai Hospital in Toronto, calls the intracellular tyrosine kinase Fps the "Rodney Dangerfield of protein kinases." Just because its cousin Src is built from a gene victimized by the first-known animal tumor virus, Src gets all the respect—or the *src* gene does anyway. Signaling researchers like Pawson think Fps deserves celebrity status, too. Not just because its gene can also be corrupted by a tumor virus, or because an analysis of what went wrong shed light on how such viruses could turn an ordinary cellular protein into a lethal weapon, but because, in addition, Fps led Pawson to the discovery of one of the most popular interaction domains in eukaryotic signaling pathways.

Pawson discovered that the Fps kinase of healthy cells (encoded by an intact gene, designated "c-*fps*" to differentiate it from the version damaged by the virus, designated "v-*fps*") has two domains: at one end of the protein, a catalytic domain responsible for manipulating phosphate, and at the other end, a domain essential to the protein's descent into cancerous delinquency. It was this second domain that was targeted by the virus, giving rise to the mutation that transformed c-*fps* into v-*fps*. What's more, Pawson learned, the cancer-causing domain was not peculiar to Fps but was a feature of all tyrosine kinases that resided in the cytoplasm instead of the plasma membrane—including Src, where it seemed to play a crucial role in making this kinase mind its manners. Intact Src speaks only when spoken to. The rest of the time it's silent because the critical domain binds to a phosphorylated tyrosine on the other side of the protein, blocking access to the enzyme's active site. "The kinase is rolled up like a pillbug," says Pawson. "It bites its own tail, causing a conformational change that inactivates it." Only when an extracellular signal breaks the bond does Src unwind and start talking about growth and division. Src-gone-bad, on the other hand, literally can't keep its mouth shut. The tumor virus that mauls it has deleted the phosphorylated tyrosine; without a handhold, the interaction domain cannot hold the protein together, the untethered kinase can babble nonstop, and the poor cell is deluded into thinking it's tapped into a bottomless pool of some growth factor. Poor Fps—overshadowed again. Pawson called the dangling interaction domain "SH2," for "Src-homology 2."*

The receptors for many growth factors are themselves tyrosine kinases, with a message for transcription factors that mediate the expression of genes critical to cell proliferation and maturation. Access to these regulatory proteins, sequestered in the cell nucleus, is tightly controlled by a sequence of phosphorylate-and-activate reactions catalyzed by a trio of kinases. Scientists call them "MAP—for 'mitogen-activated protein'—kinases," but you could think of them as the Three Fates of Signal Transduction because they control the destiny of so many signals and cells. Playing the role of Clotho, the Initiator, is the MAPKKK, short for MAP kinase kinase kinase. She phosphorylates, and activates, the MAP kinase kinase, or MAPKK. The picture of continuity, what this enzyme receives from one she

*Another molecular biologist, Saburo Hanafusa, working independently of Pawson, identified the same SH2 interaction domain in cytoplasmic tyrosine kinases. In addition, Hanafusa described a second Src-related interaction domain he called "SH3."

gives to the next, bequeathing phosphate and message to her sister the MAP kinase (MAPK), who completes the sequence and intercedes between the world outside the nucleus and the sacred space within.

Naturally, there are relays to connect growth factor receptors and the MAP kinases, pathways that converge on a GTPase known as Ras. And therein lies a problem—the receptor can neither stretch far enough to reach Ras, nor force the eviction of GDP to activate it. The receptor would be reduced to talking to itself if not for the intervention of an adaptor protein and a nucleotide-juggling catalyst, made possible, Pawson discovered, by *src*-homology interaction domains.

The receptors are a conspiratorial lot; after they bind a growth factor, they pair off and phosphorylate each other's tyrosine residues. That makes an SH2 domain the perfect confidant. Lucky for the receptors, one is waiting nearby, part of the adaptor protein Grb2; remember it as "*grab* two," because that's what it does: with one hand—the SH2 domain—it clasps a phosphorylated receptor tyrosine residue; with the other—an SH3 interaction domain—it clasps SOS, the protein that masterminds the substitution of GTP for GDP on the dormant Ras GTPase. And when the message arrives at the MAP kinases (after another hand off or two), interaction domains once again keep the conversation going. A scaffolding protein offers a safe haven where MAP kinases can sit down and talk, passing information smoothly from first to last.

Pawson compares interaction domains like SH2 and SH3 that link signaling proteins to the studs on the surface of Lego bricks. Like Lego bricks, proteins fitted with these domains can be snapped together to form sentences of any length. If signaling elements are far apart, adaptor proteins sporting multiple interaction domains can span the distance to connect them, their specificity ensuring that the word order is correct. From an evolutionary point of view, domains are economical as well. If a new demand creates a need to recast a sentence, proteins can be shifted from one signaling pathway

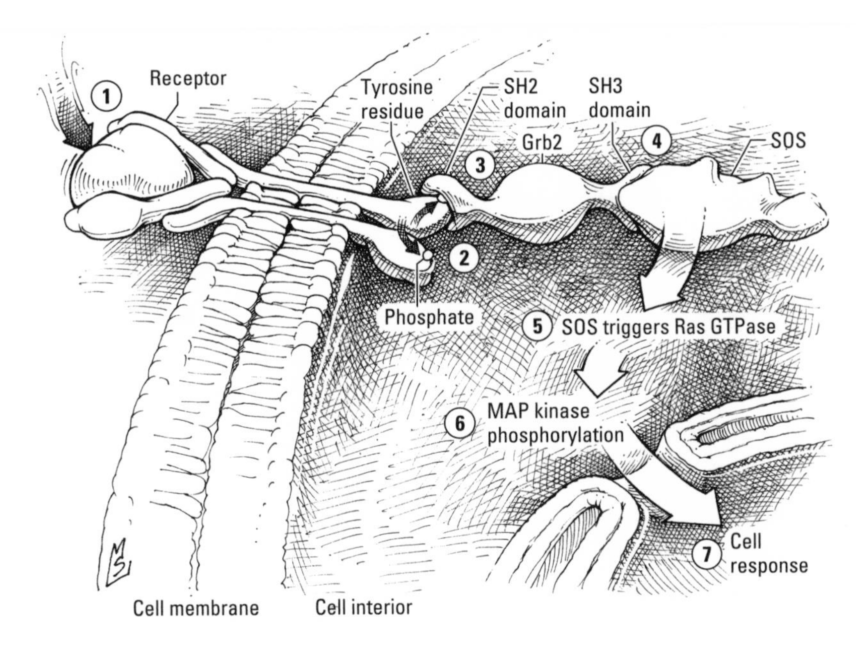

A generic receptor tyrosine kinase. The binding of a signal—a growth factor, for example (*1*)—triggers the pairbonding of two of these receptors, enabling each to phosphorylate tyrosine residues on the cytoplasmic tail of its partner (*2*). The Grb2 adaptor protein binds to the phosphorylated tyrosine of the receptor by way of an SH2 interaction domain (*3*) and to the next protein in the signaling sequence, SOS, by way of an SH3 domain (*4*). SOS then flips the Ras GTPase switch to the "on" position (*5*) and activates the MAP kinases (*6*), which transmit the message to the cell nucleus (*7*).

to another simply by splicing an interaction domain to create a new connector, rather than waiting around for the evolution of several entirely new proteins. "Interaction domains allow for rapid evolution," says Pawson. "Say you invented a tyrosine kinase and wanted to use it to signal diverse targets—to interact with or modify maybe 100 different proteins. You would need to invent different ways of doing it for every target, an option that would take lots of evolutionary time. Whereas if you invented an SH2 domain to bind phosphorylated proteins, you would only need to do it once. Then you could take that domain, stick it into a protein, and kerblam!—you can interact with the receptor."

Leave it to resourceful eukaryotes to find a way to protect a cell's hearing and get the most out of a receptor protein at the same time. Repetition isn't any more informative to a liver cell getting its ear talked off by a torrent of adrenaline than it is to a bacterium swimming in sugar, and so the muse that inspired the use of methyl groups to fine-tune the sensitivity of bacterial chemotaxis receptors came up with the idea of using phosphate groups to regulate β-adrenergic receptor sensitivity. Desensitization-by-phosphate can be delegated to one of two kinases: protein kinase A—the same PKA that activates phosphorylase kinase in the "free glucose" campaign—or an enzyme developed specifically for the job, a G protein-coupled receptor kinase, or "GRK," recruited by the Gβ-Gγ duo in retaliation for their expulsion by Gα. It's rumor-mongering, protein-style; regardless of which kinase does the deed, the β-adrenergic receptor's romance with G_s is history.

Worse, a receptor time-out can be a prelude to exile or even execution. GRK continues its meddling by introducing the now unattached and phosphorylated β-receptor to a new suitor, β-arrestin. A scheming rapscallion of a protein, β-arrestin is involved in some shady goings-on with certain membrane proteins—machinations that typically end in the receptor's abduction; surrounded, then spirited away inside the cell in a pinched-off bleb of membrane. If the receptor's lucky, it will escape with its life and eventually find its way back to the plasma membrane. Should β-arrestin's mood turn surly, however, the receptor will be abandoned to certain death in its membrane-bound prison, dissolved by a brew of acidic chemicals and vicious protein-eating enzymes.

But that's wasteful. Why discard a perfectly serviceable receptor when you can redeploy it, arranging a transfer to another pathway that can still make good use of it? PKA does that—rather than sending the receptors it phosphorylates to the trash bin, it turns them over to the care of another G protein. Seconded to G_i instead of G_s, the β-receptor's new job description calls for a rendezvous with the

MAP kinases, as if it were a growth factor receptor instead of a hormone receptor.

β-arrestin can direct the receptor to a new career in growth regulation as well, if it's dressed up as an adaptor. When β-arrestin-the-connector comes to call on a phosphorylated β-adrenergic receptor at the membrane, it comes with three MAP kinases in tow, clinging like burrs to interaction domains built into its topography. Gone is the old preoccupation with catastrophe. Instead of being deported or destroyed, the β-adrenergic receptor gets a new lease on life—and the cell gains another new way to discuss matters of growth and division, without inventing a single new protein.

Architect Christopher Alexander puts it simply: "Patterns," he notes, "need the context of others to make sense." Adaptor and scaffolding proteins incorporating versatile interaction domains allowed eukaryotic cells to join protein to protein and pattern to pattern, mixing and matching components to create novel pathways and larger networks. In doing so, they added context without sacrificing clarity, increasing the versatility of signaling proteins while obviating the need for a supercomputer-sized genome.

A LIVING LANGUAGE

Begin your building project, suggests Alexander, with a foundation that doesn't just sit on the ground but is actually connected to it; sink columns, like tree roots, into the earth below to improve stability and to integrate the building with its natural surroundings. Join column to column at the top with perimeter beams to frame rooms and build a base of support for the next story. Position windows and doorways to take advantage of natural light, highlight a beautiful view, facilitate the flow of traffic between rooms; insert half walls and alcoves to separate activities and afford privacy without isolating people behind closed doors. If possible, break up a single building into several units—a cluster of small shops or a house and an adjoining cottage, keeping at least half the site as open space. Finally, don't be selfish and shortsighted—consider your project an opportunity to improve the community at large, locating it to minimize sprawl, preserve green space, and maintain an optimal population density.

According to Alexander, the hierarchal nature of an architectural "pattern language" means that the solutions to the largest design problems—How large should a city be? How can we achieve a balance between urban, suburban, and rural areas? How do we design communities that work toward common goals yet respect diversity?—grow naturally out of the solutions to the smallest—floors and walls, windows and doorways. In such a hierarchical system, every pattern "helps to complete those larger patterns which are 'above' it, and is itself completed by those smaller patterns which are 'below' it." A builder need never start entirely from scratch; rather, using a common set of design elements and integrating those elements to form larger patterns, he or she can generate any house, every neighborhood, even an entity as large as a city or a community.

Nature understands this "timeless way of building," homing in on adaptive solutions to fundamental biological problems, then elaborating on these patterns. As developmental biologists John Gerhart and Marc Kirschner note in their book, *Cells, Embryos, and Evolution*, "where we most expect to find variation, we find conservation, a lack of change. There is divergence in the accumulated genetic changes . . . and in protein sequences, but there is conservation in the function and structure of many of the mechanisms of the cell." Rather than demolish and rebuild, they argue, cells are more likely to recycle, to change the way existing mechanisms are regulated or utilized, rather than the way they work.

Eukaryotic signaling mechanisms exemplify this principle of conservation. Tried and true solutions to the fundamental problems of cell-cell communication, the basic design elements that constitute the language of life, pioneered by simple organisms, were "carried forward in modern, more complex organisms, elaborated upon but not really changed all that much," says Kirschner. "So they must have been a very good framework on which to build a very complex organization of a large number of cells, with a lot of cell differentiation and spatial organization."

Yet paradoxically, signal transduction mechanisms, paragons of conservation, have also been agents of great change. The bridge between the social environment that surrounds the cell and the internal machinery that frames its responses, signaling pathways are respected members of the management team regulating core functions. As a result, changes in "contingency" based on the transfer of responsibility for a conserved function from one signaling pathway to another have become one of the most popular ways for multicellular eukaryotes to experiment without endangering essential processes, going hand in hand with the evolution of these organisms.

G protein-coupled receptors, one of the great linguistic innovations of eukaryotic organisms, demonstrate brilliantly how eukaryotic signaling mechanisms manage to be both conservative and flexible. The receptor protein itself is the epitome of consistency. Like other transmembrane receptors, it comprises an extracellular binding site, a unique configuration of ridges and clefts that correspond, lock-and-key fashion, to the three-dimensional structure of the signal; a midsection submerged in the plasma membrane; and a cytoplasmic segment charged with activating an intracellular signaling relay. It always has the same seven-helical structure. And it always gets its point across by doing a little dance with a G protein, a molecular switch toggled on and off by the binding and breakdown of GTP.

That two-protein two-step introduces plenty of opportunities for improvisation, however. Through the clever use of kinases and adaptors, the β-adrenergic receptor, for example, can be dancing with G_s one minute and then taking a break, preparing to die, or waltzing off with a stranger the next. Adrenaline's message is usually a call to arms, but when a sentence typically headed by a growth factor receptor kinase is rewritten this way with a G protein-coupled receptor as its subject, it changes its tune, placing cell growth under the aegis of a new contingency.

The modular construction of G proteins also fosters creativity. G_s, G_i, and all the others, like the receptors they assist, have a common structure: Gα, the anchor; Gβ, the propeller; Gγ, the tail. Within that framework, however, cells have many options. Sixteen different genes code for the α subunit of the tripartite G protein. β subunits are encoded by five. Twelve genes specify γ subunits. By mixing and matching α, β, and γ genes, a cell can create, in theory at least, dozens of different G proteins. And when you add over 800 options for your receptor, "you get a lot of flexibility," says Al Gilman. "Each cell can go to the genome as if it were going to the local Radio Shack. It can shop for the kind of receptor it needs today, with hundreds to choose from. Then it can go to the counter and start picking a G protein, again, with dozens of possibilities. By combining components, it can build its own custom switchboard, adapted to meet its own individual needs." Simply by changing receptor-G protein combinations—changing contingencies—an organism can redeploy a conserved process in different types of cells, under different circumstances, or at different times in its life.

Skillful use of mobile interaction domains offered yet another way to conserve patterns while manipulating contingencies. Easily spliced into a new gene, archetypal interaction domains like SH2 and SH3 transformed ordinary proteins into versatile Lego bricks, equipped for jobs in signaling pathways as adaptors or scaffolds. Receptors and kinases continued to do the heavy lifting, while proteins fitted with these interaction domains worried about details like word order, recruiting and assembling signaling components in an approved fashion.

But adaptor and scaffolding proteins did more than prevent verbal anarchy. Domains introduced a new level of complexity, enabling cells to merge signaling pathways to form networks and redirect components to other partners in alternative pathways. A linguist would say that they made the language of life more productive, meaning that eukaryotic organisms were not restricted to a handful of truisms but could combine signaling proteins in innovative ways as they evolved, generating novel sentences as needed to guide the behavior of large numbers of cells in a wide variety of circumstances. One of the features that distinguish spoken language from the vocalizations of animals or the rigid, formal "languages" of computer science, this flexibility has given multicellular eukaryotes a breathtaking capacity for innovation, a freedom to play with the identity, lifestyle, and spatial organization of cells that has been critical, Gerhart and Kirschner conclude, to the diversity of these organisms.

"Patterns have enormous power and depth; they have the power to create an almost endless variety," Alexander concludes. Starting with walls and roofs, doors and windows; building up to houses and combining them with offices and stores; adding parks and highways, people can build a city from simple design elements. Similarly, combining signals and receptors, second messengers and G proteins, teams of kinases, adaptors and scaffolds, and linking pathways to create networks, "a few thousand gene products can control the sophisticated behaviors of many different cell types." Just as the smallest patterns help to complete the larger patterns of neighborhood and city, these smaller molecular patterns help complete complex signaling pathways that, in turn, help complete the even larger patterns of cell and organism.

Doors slam and dogs bark—Jenny and Haley are home from school. Backpacks thump. Someone ruffles through today's mail; someone else drops a CD player. A quarrel about the ownership of a hat is already taking shape.

"Holla!" I venture.

"Oh my God," they reply in the tone reserved for my most egregious displays of parental simplemindedness. "NO ONE says that anymore."

Don't worry—they will have new words for me to learn. The adoption of words from pop culture, changes in meaning and pronunciation, the relentless influx of slang, the influence of technology and globalization continue unabated, living proof that language is a living entity. It never stops growing or changing, and, if you listen carefully to the younger generation, you learn something new every day.

PLAITING THE NET

Have you ever thought, while you're struggling to balance a bag of groceries, the dry cleaning, the videos your kids forgot to return yesterday, a two-liter bottle of diet soda, and the car door that it would be really, really helpful to have an extra arm or two? If only you had realized how easy it would have been to create a limb when you were an inch-long embryo—all you would have needed were the right words.

Just in case you ever manage to turn back time, those words are "fibroblast growth factors."

Vertebrate limbs begin as little hillocks of undifferentiated tissue known as "limb buds" that erupt at precise spots along the lateral edge of the embryo. More than 50 years ago, embryologists analyzing the development of legs and wings in the chick embryo discovered that the outgrowth of these limb buds was masterminded by a lip of tissue curving around the leading edge of the bud, the apical ectodermal ridge: If the ridge was cut away, the limb stopped growing; if it was rotated 90 degrees to the left or right, the limb grew outward from the body at right angles to its normal orientation. But it wasn't until the 1990s that scientists learned that fibroblast growth factor, or "FGF," was what the ridge said to the bud. A plastic bead impregnated with FGF and apposed to the stump after amputation of the apical ectodermal ridge could substitute for the missing tissue, spearheading the development of a complete wing or leg indistinguishable from a normal limb. What's more, researchers learned, a little FGF on a bead could not only rescue an established limb bud, it could also make an extra limb sprout in the space between wing and leg. "You get a complete limb," says developmental biologist Cliff Tabin, "a humerus, a radius and ulna, a wrist, and digits."

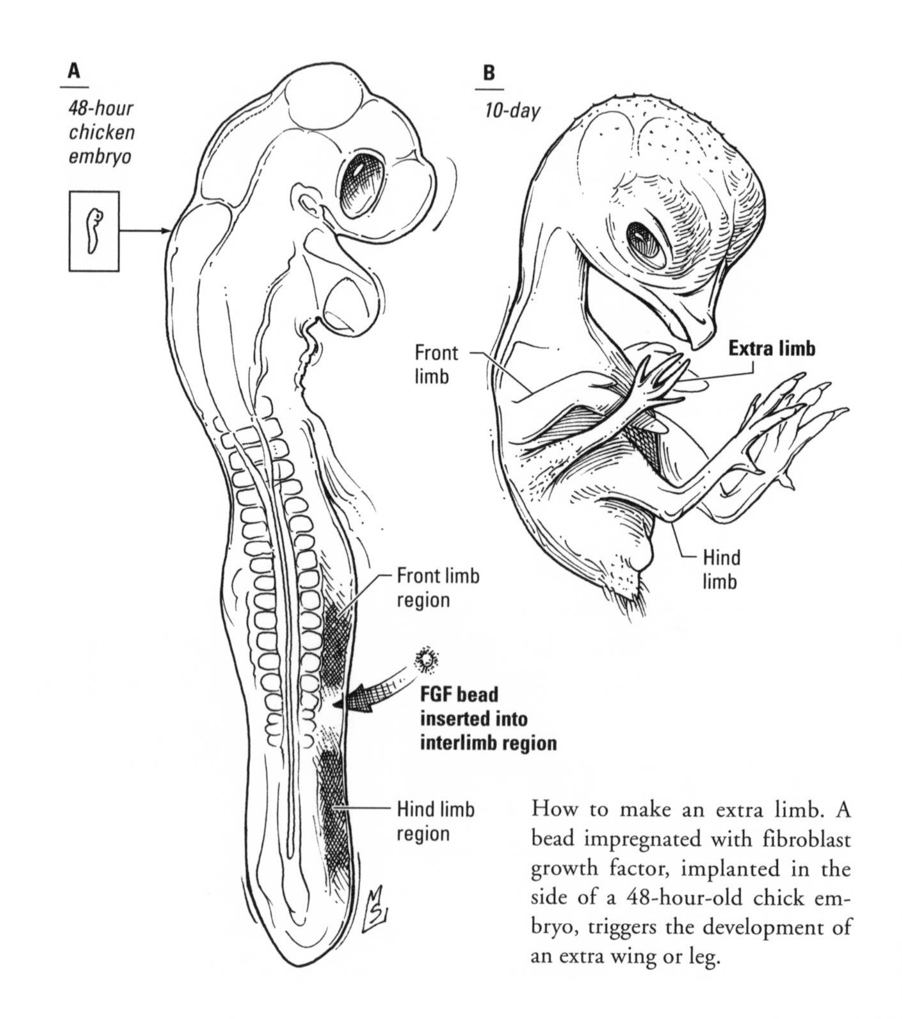

How to make an extra limb. A bead impregnated with fibroblast growth factor, implanted in the side of a 48-hour-old chick embryo, triggers the development of an extra wing or leg.

Prefer something a little more realistic? No problem—a pellet of FGF-secreting cells can be substituted for the bead. And don't be concerned because you aren't a chicken. So-called knockout mice, created by scientists wielding genetically engineered weapons that allow them to take potshots at specific genes, have shown that FGF is as important to limb development in mammals as it is in birds. Provided you'd taken care to apply FGF before your two normal arms began to develop, you would have been on your way to increased dexterity without a second thought.

Perhaps, if you have young children, you could use eyes in the back of your head, as well as an extra arm. In that case, you should have added "bone morphogenetic protein number four," or "BMP4," to your vocabulary. In collaboration with FGF, BMP4, secreted by the optic vesicle (a pedestal of tissue protruding from the embryonic brain), talks nearby cells into fashioning the eye's lens; signals issued by the lens, in turn, instruct the optic vesicle to make a retina. If you had grafted a couple of these BMP4-secreting optic vesicles under your scalp, today you'd be able to keep an eye on what's happening in the rear seat without ever taking your eyes off the road. Maybe you're a pianist or a writer, and a few additional fingers would be more useful than another arm or a second pair of eyes. If so, the phrase you needed to know was "Sonic hedgehog." A pellet of cells producing this signaling molecule, implanted in the head of the limb bud, would have given you a duplicate set of fingers. Or maybe you prefer brains to brawn. Provided you don't mind people staring at you as if you had two heads—because you *would* have two heads—your word would have been "Cerberus." With the correct words ("Noggin" and "Chordin") and good timing, a truly narcissistic individual could even generate a Siamese twin—given, that is, the individual in question was a narcissistic amphibian.

Arm or eye, animal or human, house or cathedral, a building project of any complexity is necessarily a collaborative effort. As art historian J. J. Coulton puts it, "Whereas a statue can in most cases be completed by one man with his own hands, this is normally impossible in architecture." As a result, Coulton observes, the evolution of architecture in ancient civilizations could not have occurred without the evolution of techniques for describing buildings and their construction, "to communicate the architect's intention to the builders." For example, builders in Egypt and Mesopotamia carved outlines of proposed structures on clay tablets or sketched them on sheets of papyrus—the earliest known examples of the architectural ground plan. Egyptian architects also pioneered the use of front- and side-view elevation drawings. What's more, these drawings were rendered on papyrus divided into squares, evidence that their creators were already familiar with the square grid as a technique to specify proportions.

Metazoans embarked on a building program as ambitious as that undertaken by any ancient civilization and as equally dependent on effective communication. Much as the progression from tents to pyramids went hand in hand with the invention of the floor plan and the square grid, the progression from single-celled organism to a multicellular body correlated with the evolution of mechanisms for stipulating the spatial relationships between specialized tissues dedicated to the administration of distinct physiological functions. In place of brush or stylus, multicellular organisms had only chemicals; using no more than the pattern language of design elements so valuable in other contexts, they were expected to discriminate head from tail, mark the placement of limbs, ensure that the eyes were in the head and the feet on the ground.

But these enterprising creatures faced a technical challenge more formidable than any confronting their human counterparts. An architect undertaking the construction of a temple or palace began with stacks of bricks, yards of timber, and legions of slaves. A metazoan embryo began with a single fertilized egg.

"*Ex ovo omnia*"—all come from eggs—physiologist William Harvey declared in 1651. But how did a large, complex creature emerge from such humble origins? The naturalists and philosophers of ancient Greece were among the first to search for an answer to this riddle. Some studied birds' eggs and the fetuses of domestic animals and declared that a fully formed creature was present from the very beginning, only much, much smaller. For these "preformationists," development was about growing larger, a grace period in which the youngster took shelter in the egg or the womb until it was big enough to survive in the outside world. Others, led by Aristotle, examined the same materials and reached a totally different conclusion. They claimed that the embryo began not as a miniature adult but as an undifferentiated mass. Structure and organization emerged gradually, "as we read in the poems of Orpheus, where he says that the process by which an animal is formed resembles the plaiting of a net." Epigenesis, as this viewpoint came to be called, held that development was a time of differentiation and maturation in which the organism grew in complexity as well as size.

"If organic form is not original, but is produced, what accounts for regularity and directedness of process? And how do all individuals of the species end up with the same body plan, the same organization of the internal organs, the same proportions of each type of cell?" Epigenesis, with its insistence that living things were constructed rather than created, raised uncomfortable questions about the nature of development. What mysterious force or invisible hand guided the embryo's journey from nonentity to individual? The shadow of heresy was not enough to deter Harvey, but most of his contemporaries were slavish devotees of preformation. Pious sorts clung to preformation because it could be reconciled with their belief in a Creator-God. And the enlightened embraced it because it appealed to their belief in an orderly universe; they saw, in the systematic appearance of generation upon generation of prefigured organisms, a scheme as law abiding as the movement of the stars or the actions of gravity.

You might think that the introduction of the microscope would have settled the argument. It did not. If "seeing is believing," most still wanted to believe in preformation, and so they convinced themselves that they saw little men and tiny animals when they peered into their microscopes.

Others finally dared to believe their eyes. Let those blinded by long-standing prejudices defend the faith. "What [one] does not see is not here," insisted Caspar Friedrich Wolff, one the first of these revisionists, in 1759. Gradually and grudgingly, even the most resolute had to agree. But while the wholesale conversion to epigenesis put an end to fictitious accounts of miniature organisms, it brought biologists face to face once again with the old question of how the egg gave rise to a fully formed organism.

When preformation was the order of the day, embryologists had the easiest job in biology—if the embryo did nothing but grow, the study of development required no more than watching and waiting. But the mechanisms underlying epigenesis could not be deduced by observation alone. To solve this problem, scientists had to use their hands as well as their eyes. Wielding needles, scalpels, and dyes, they turned developmental biology, long a spectator sport, into an experimental science.

In 1901, Hans Spemann, a lecturer in zoology at the University of Würzburg and one of the new evangelists of the experiment, published a paper explaining how the tadpole got its eyes. In the frog *Rana fusca*, as in other vertebrates, development of the eye begins with the optic vesicles, situated on either side of the primordial brain. As the vesicles expand, they contact the overlying tissue, or ectoderm, retract, and differentiate to form the retina, while the ectoderm itself pulls away and divides to fill the cup of the developing retina; this filling becomes the lens. Spemann suspected that the vesicle somehow influenced lens development; to test his hypothesis, he used a scalpel made from a glass needle and a loop of baby's hair to pinion the tiny embryo and excise, reposition, or exchange slivers of tissue.

Spemann reported that when he removed the entire optic vesicle early in development, neither eye nor lens developed on that side of the embryo. But if he left even a remnant of vesicle tissue in place, next to the ectoderm, an extraordinary thing happened—despite the absence of a complimentary retina, the ectoderm still developed into a lens. Furthermore, lens making was a secret peculiar to the optic vesicle. Other tissues, transplanted to the head, could not make a lens. But if the vesicle was cut free of the brain and moved so that it contacted the ectoderm of the trunk instead, why, a lens formed there, in the belly of the tadpole! Spemann concluded that lens development was directed by a very private and specific conversation, in which the optic vesicle induced the production of a lens through some sort of exchange with the receptive ectoderm.

The most obvious way for one tissue to communicate with another was by touch. At first that was the way scientists like Spemann thought embryonic induction occurred. But in 1952 another researcher—a mathematician, not an embryologist—suggested that transactions between tissues during development might also be couched in the language of chemistry. In a paper entitled "The Chemical Basis of Morphogenesis," Alan Turing (better known as one of the pioneers of computer science) proposed a model "by which the genes of a zygote may determine the anatomical structure of the resulting organism," based on the hypothetical interactions between two substances he called "morphogens" ("form-giving molecules"). Turing's "reaction-diffusion" model posited that these substances, when mixed together, diffuse freely and interact in two different ways. Morphogen A, which diffused more slowly than Morphogen B, was a catalyst; it enhanced not only its own production but that of Morphogen B as well. Morphogen B, on the other hand, diffused rapidly and inhibited the production of both substances. Turing's mathematical analysis revealed that under certain conditions the interaction of Morphogen A with Morphogen B could generate recurrent spatial patterns, peaks and troughs of morphogen concentration similar to the regularly spaced patterns of leaves, petals, tentacles, and stripes found in plants and animals—suggesting that the distribution of chemicals could indeed draw a "floor plan" that cells might translate into biological structure.

Floor plans and elevations gave architects of antiquity what Coulton calls "a technique of design," a solution to the problem of how to communicate the placement of doors, the spacing of columns, or the dimensions of rooms to builders. Similarly, patterns based on the spatial localization of chemical signals represented a "technique of design" that metazoan organisms could use to solve such problems as the placement of organs, the shape of the body, and the differentiation of specialized cells. In these patterns the developing embryo could discern the fates of cells and extract the information needed to craft an entire body, step by step, from a single cell.

MEET THE CONTRACTORS

A completed arm is an alliance of nerve and muscle, skin and bone, tendons and joints. But what makes a nerve cell nervous or a muscle cell muscular? In the cellular world, personality is based on proteins. Each type of cell is characterized by the unique constellation of proteins it synthesizes: neurons make ion channels and pumps that control the concentrations of ions and the propagation of electrical impulses, muscle cells string cables of protein fibers and mount protein motors to crank them taut, red blood cells pack themselves with hemoglobin for carrying oxygen, cells lining the gut produce enzymes to break down food. Each elaborates its own protein repertoire, yet all came from an egg containing the only copy of the organism's genes.

When embryology was a young science and developmental genetics had yet to become a recognized discipline, one popular explanation for this paradox held that the genetic material bequeathed to the egg was divided among the daughter cells in the early rounds of cell division. A cell's fate was determined by its particular allotment:

cells that received muscle determinants gave rise to muscle cells, those that received determinants peculiar to the skin gave rise to skin cells, and so on. In modern developmental biology, the Great Partition has been superseded by the Great Decision. During development, we now know, the genome is copied and forwarded in its entirety, at every cell division, to all progeny of the original egg, not split up and parceled out. Each cell, however, actually uses only a fraction of the thousands of genes held in common. As it progresses from egg to embryo to newborn, the cell can alter its gene selections in response to external cues, including signals issued by its neighbors. At maturity a cell's final identity is a summary of the decisions it has made regarding gene expression, culminating in the selection of a battery of genes encoding the proteins needed to perform a single specialized task.

"DNA makes RNA makes protein." That's the mantra you may have memorized in high school biology, but in the living cell there are a lot of "ifs," "ands," and "buts" in between. These contingencies enable cells to use the same genome in a multitude of ways, decisions that chart the course of embryonic development.

Manipulating the expression of genes in eukaryotic cells is so easy because transcribing eukaryotic genes is so hard. Before a gene can be transcribed at all, it must have permission. Even then, RNA polymerase, the enzyme that actually does the work, is baffled by the way DNA is twisted, scrunched, and tied up in protein in order to pack it into the chromosomes; is all thumbs when it comes to unraveling the double helix; and is clueless about the right way to straddle a strand once it's unwound. If cells had to wait for the polymerase to figure out transcription on its own, the job might never get done.

To make proteins, DNA needs proteins, a corps of mechanics and decision makers that biologists call "transcription factors." These proteins come in two flavors. So-called general transcription factors assemble at the promoter sequence where transcription begins; they recruit the polymerase and orient it correctly. Other transcription factors are activators and repressors that permit or forbid gene expression. They bind to a distinct regulatory sequence that is not part of the gene itself and may, in fact, be located a considerable distance upstream or downstream. Gene activators give a thumbs up to RNA polymerase, recruit and organize the enzyme's crew of general transcription factors, and exhort the whole conglomerate to work harder and faster. Repressors stifle gene expression by blocking the binding of activators, interfering with their recruiting efforts, or smothering the DNA in more protein.

Both activators and repressors recognize and bind to DNA regulatory sequences much as receptors recognize and bind signaling molecules or interaction domains detect consensus sequences. If the idea seems surprising that's probably because you're used to thinking of the DNA double helix as a ladder. In reality it's a spiral staircase of rough-hewn stones—its constituent bases, each outlined in a characteristic pattern of chemical features more or less eager to form transient bonds with passersby. Collectively, these features generate a surface as unique as that of any protein. Just as scaffolding proteins contain interaction domains that correspond to consensus sequences, transcription factors have evolved structural motifs that "read" DNA sequences by interacting with the surface features of the bases contained in that sequence. Some sport a coil, or a loop closed with an atom of zinc that fits into a groove of the helix. Some contain a strip of adhesive amino acids that latch on to their cognate sequences like Velcro. Others grip the DNA like a clothespin or join hands with a second gene regulatory protein and encircle it.

In a classic series of experiments carried out in the 1950s, François Jacob and Jacques Monod revealed how transcription factors enable even an organism as unpretentious as *E. coli* to use the information stored in its genome selectively. The well-prepared bacterium has genes for enzymes that allow it to metabolize two sugars: glucose, its preferred food, and lactose. In addition, it has two regulatory proteins that allow it to change its complement of metabolic enzymes to suit today's menu. When lactose is unavailable—or glucose is plentiful—Jacob and Monod found that an inhibitory transcription factor, the lac repressor, shuts down transcription of the gene that encodes the lactose-processing enzyme, conserving time, effort, and raw materials that would otherwise be invested in producing an unnecessary protein. But when lactose is the only lunch option, *E. coli* substitutes an activator called CAP for the repressor, turning on the enzyme gene so that it can digest this sugar. No matter what the environment serves up, by micromanaging its genes, *E. coli* can endeavor to stay well fed while husbanding energy and resources.

Many eukaryotic genes seem to have felt transcription was too important to be entrusted to a single regulatory protein. Their expression requires the combined input of an entire consortium of transcription factors. The imposition of additional constraints not only afforded eukaryotic cells an exquisite degree of control over gene activity but also offered unlimited opportunities to customize gene expression at different times and in different places, merely by increasing or decreasing levels of the gene regulatory proteins that collectively determine whether the gene is turned on or switched off. In addition, by integrating transcription factors into signaling pathways, gene expression could be placed under the aegis of external signals, granting a cell's neighbors the power to induce long-standing changes in its behavior, its activity, or even its identity.

Although it is often compared to a blueprint, the genome is more like a home design magazine, full of clever ideas, new products, amazing gadgets, and grand decorating schemes—only some of which are useful. Contractors and homeowners, paging through these options, must select the features best suited to the task at hand. All-white cabinets? Impractical with toddlers. Shiny chrome fixtures? Too old-fashioned—brushed nickel is what everyone's choosing these days. A prefab shower won't fit the space; the hand-painted ceramic tile won't fit the budget.

Embryonic cells, committed to a building enterprise that makes a kitchen renovation look like a child's craft project, also have choices to make—choices about genes. The dynamic control of gene expression via transcription factors and signaling pathways facilitates these decisions. During development, each cell's contractors and homeowners—transcription factors and the extracellular signals that boss them—page through thousands of options to select the right genes, first to frame and then to finish the organism. Just as the configuration of the plumbing determines the placement of the sink, the size and shape of the sink affect cabinet selection, the finish on the cabinets determines the best tile color, and the amount still left in the bank dictates whether it will be paint or paper, the decisions cells make at one phase of development determine their options in the next phase. Today's effect becomes tomorrow's cause, and successive rounds of gene expression gradually partition the embryo into smaller and smaller populations of like-minded and increasingly specialized cells, each characterized by a distinctive pattern of gene activity and, as a consequence, a distinct repertoire of proteins. Manipulating genes in sequence, neighbor guides neighbor as the embryo maps out the rudiments of a body plan, then builds on that foundation to assemble a mature organism.

THIS END UP

Writers are derailed by writer's block, musicians and actors by stage fright, but if any group of artists has a right to be paralyzed by fear of failure, it's architects. As Coulton observes:

> A man modeling a clay figure can . . . to a certain extent modify the part formed first in the light of what he does next, and can even reject the whole form and start over again without much loss; a painter is in much the same position. The sculptor working in marble is rather more constrained, for a serious mistake will be both irremediable and costly, but he can work gradually into the stone over the whole of his figure, so that the relation of the parts to each

> other can be clearly visualized at a stage where minor changes to any of them are still possible. The architect on the other hand must always start his buildings at the bottom and cannot modify at all what he has built first in light of what follows. Mistakes made at the start can therefore not be corrected, and they will also be ruinously expensive.

A developing embryo also must start at the bottom, cannot look back, and pays a heavy price—death or disfigurement—for mistakes. To "ensure that the lower parts of the building will suit the parts to be put upon them," the embryo's floor plan must answer fundamental questions about orientation and layout, or its project will collapse before it ever sees the light of day.

Specifically, developmental biologist Marc Kirschner says, the nascent embryo must accomplish three critical tasks. First and foremost, it must acquire a sense of direction. "The first thing you need is some global organization, some global polarity—head-to-toe polarity, back-to-front polarity. If you had a lot of autonomy, you could easily imagine an organism sprouting multiple heads. So early on in the process of development, you have to have some sort of way of generating a polarity in the organism that will establish the early anatomy of the organism, will absolutely dictate that anatomy and not allow for deviations from that," he says. Biologist John Tyler Bonner goes even further. "Perhaps the most fundamental property of any development—including the development of unicellular organisms—is polarity," he writes, a feature so basic, so ancient, it can be considered a "kind of living developmental fossil."

"The next thing would be an inside and an outside," Kirschner continues. Sealing off the interior is an essential prelude to the acquisition of a "capacity for reliable signaling within the organism," as Kirschner calls it, the faculty that enables the embryo to realize complete mastery over the behavior of its constituent cells. To achieve this goal, the clump of cells formed during the first cell divisions must be drawn apart, rearranged, or carved out to form a hollow sphere, in which some cells face outward and bind together to form a secure barrier—an epithelium—while the rest retreat inward. The fastidiously composed fluids contained within the epithelium will constitute their reality, the signals dispersed within it, the voices that will guide them for the rest of their lives.

Finally, Kirschner argues, the embryo must mark out a system of compartments that will be the basis for all subsequent differentiation; having laid the foundation, it must frame out the rooms, as it were, by manipulating gene expression to subdivide the front-to-back and top-to-bottom axes specified in the first step. Embryos, he says, need "mechanisms to set up an 'invisible anatomy' of signaling pathways, often in a sort of segmental pattern, anterior to posterior, and another invisible anatomy to set up an invisible pattern from dorsal to ventral." For example, patterning mechanisms might mark out 20 segments from head to tail and 10 divisions from back to front, a total of 200 individual domains. Within each of these compartments, cells share a common bond—a pattern of gene expression found only in that compartment and no others. "This allows the first major kind of modularity," Kirschner concludes. "All those cells, they may develop into things very different within those domains, but they're different from every other domain in that they have that special address"—and that common pattern of gene expression.

Polarity, insularity, modularity—these three properties abide in metazoan embryos. But the greatest of them is polarity.

Instant polarity—just add water. To set a course, an embryo must have a compass, and morphogens—the diffusible, form-giving substances proposed by Alan Turing—suggest one way to point the youngster in the right direction. The embryo can simply take advantage of the fact that as a diffusible chemical flows from an area of high concentration to an area of low concentration, it automatically creates "a directional arrow that points down the slope of the [concentration] gradient." To learn how far back or how close to the top they are, cells "read" the concentration gradient; based on this information, they activate a corresponding battery of genes, translating the polarized spatial pattern formed by the chemical into a molecular pattern capable of generating a spatially organized body.

To illustrate this principle, developmental biologist Lewis Wolpert suggested how a group of patriotic cells might use "positional information" (as he called it), supplied by a morphogen gradient, to turn themselves into a replica of the French flag. Wolpert asks readers to imagine a row of cells, lined up end to end like the threads of a strip of cloth. A chemical signal, released from a reservoir on the left side of the row, diffuses from left to right, creating a concentration gradient. Each cell decides whether to activate genes for the color red, white, or blue on the basis of the concentration at its position in the gradient: cells closest to the reservoir, exposed to the highest amount of the morphogen, become blue; those farthest to the right, where morphogen levels are lowest, elect to be red; those in between, where morphogen levels are too low to turn on blue genes but too high to turn on red, become white.

On paper the morphogen gradient is a sensible candidate for an embryonic patterning mechanism; indeed, it has been one of the most pervasive models in developmental biology. In practice, identifying real morphogens has posed one of its most significant challenges. Scarce, elusive, and short lived, they slip right through the clutches of standard purification techniques. Genetics, in such cases, can often accomplish what biochemistry cannot—a rare protein that lurks in the shadows may be flushed out by a fortuitous mutation, trapped or tracked following the identification of its gene. But until the mid-1980s, every geneticist's favorite organism, the fruit fly *Drosophila melanogaster*, was hardly the ideal subject for studying the genetics of early development. Flies with odd eye colors or extra wings can survive and breed in the sheltered environment of the laboratory, but a fly embryo with a mutation in a gene critical to development—a gene, for example, that differentiates its head from its tail—is likely to die long before it ever sees the light of day. Figuring out what went wrong with a dead embryo in a tiny opaque egg frustrated even the most resolute; as developmental biologist Scott Gilbert notes, "'Death' is a difficult phenotype to analyze."

Then, in the 1980s, German embryologists Christiane Nüsslein-Volhard and Eric Wieschaus learned that a simple trick—soaking the fly's eggs in oil—rendered their impenetrable shells transparent. Even the youngest embryos could be observed in these eggs-turned-fishbowls; once their DNA was scanned for aberrations, Nüsslein-Volhard and Wieschaus could then correlate phenotype and genotype to determine the roles played by specific proteins at every stage of development.

One of the thousands of embryos Nüsslein-Volhard, Weischaus, and their colleagues described was a monstrosity with a second tail where its head and thorax ought to have been, the product of an accident involving a gene they christened "*bicoid.*" Cytoplasm from the anterior end of a normal embryo could counter the mutation, further evidence that the protein product of *bicoid* mediated the all-important task of orienting the embryo. But the *bicoid* mutation wasn't a flaw in the embryonic genome. This mistake was committed in the genome of the mother fly.

Parents everywhere want to do what they can to help their children. Detached as she may seem, the female fruit fly is no exception. She may not build a nest or fetch food, but she does make sure her offspring will know their heads from their tails, planting clues in the egg the embryo can use to jump-start the crucial axis-building enterprise. Her gifts are mRNA transcripts and they are recipes for morphogens.

The *bicoid* transcript, tethered to a complex of anchoring proteins at one end of the ovoid egg encodes one of these maternal morphogens. In the dialect of *Drosophila*, "bicoid" means "head." Translated into protein, *bicoid* yields a transcription factor that diffuses away from this end of the egg to form a concentration gradient pointing in the head-to-tail direction.* To be sure the embryo understands, the mother fly reinforces the message: in addition to *bicoid,* she has secured a complementary transcript at the other end of the egg. The protein encoded by this mRNA, Nanos, diffuses in the opposite direction, generating a gradient that points from tail to head.

In contrast to the *bicoid* and *nanos* transcripts concentrated at the front and rear of the egg, respectively, two other maternal mRNAs, *hunchback* and *caudal,* are plastered more or less evenly throughout. As Bicoid and Nanos diffuse, they fiddle with the expression of these ubiquitous transcripts. Bicoid activates *hunchback.* As a result, the Hunchback protein, another transcription factor, is actually produced only in the anterior portion of the embryo; in more posterior regions, there's no Bicoid to turn *hunchback* on. Conversely, Bicoid binds and strangles *caudal* mRNA, limiting translation of the Caudal protein to the tail end of the embryo. Nanos has it in for *hunchback.* It modifies the *hunchback* transcript and renders it indecipherable. Concentrations of Hunchback protein, therefore, plummet as soon as the transcript encounters the leading edge of the Nanos gradient. When all is said and done, the Hunchback protein defines a head-to-tail gradient that reinforces the polarity specified by Bicoid, while the Caudal protein forms a tail-to-head gradient that backs up Nanos.

The female fly decorates the egg with instructions for distinguishing its back from its belly as well. When the embryo is ready to work on this problem, about an hour and a half after fertilization, it opens and decodes Mother's last mRNA transcript, spray painting

---

*The fruit fly gets around the awkward fact that transcription factors typically could not diffuse across plasma membranes because for the first four hours of its life the only plasma membrane is the one surrounding the egg itself. The first 13 "cell" divisions after fertilization are actually nuclear divisions—after DNA replication, each nucleus divides, but no membranes form to isolate them into new cells. The result is a syncytium, a husk of free-floating nuclei surrounding a bolus of cytoplasm, in which even proteins like transcription factors can diffuse freely.

itself with a mist of Dorsal protein. Unlike Bicoid and Nanos, however, Dorsal cannot diffuse. Tethered in place to a chaperone protein, this transcription factor can't even get into the nucleus, much less manipulate genes, until it's liberated from its minder.

There—can you hear it, wafting up from the cells lining the base of the ovarian follicle? It's Mother's sweet angelic voice, crooning a lullaby that lulls the chaperone to sleep. Free of its grasp, Dorsal can gambol into nuclei. Well, some nuclei, anyway. Follicle cells have soft voices, and their song won't carry all the way across the egg. A short distance away, receptors are already straining to hear it; back up to the equator, and it's no more than a whisper; a little farther away, and it can no longer be heard at all. The fainter the signal, the less Dorsal liberated from the chaperone; the lower the concentration of free Dorsal, the less Dorsal-inspired gene expression. And the less influence Dorsal exerts, the closer that region of the embryo will be to the back of the fly.* Instead of a chemical gradient, the embryo reads a gradient of activity, created by a ventral-dorsal variation in the liberation and translocation of Dorsal, under the aegis of a ventral-to-dorsal variation in a maternal signal. Just as gradients of Bicoid and the others polarize the embryo in the head-to-tail direction, this gradient of nuclear Dorsal points the way from the bottom of the embryo to its back.

Using just a handful of transcription factors and RNA-binding proteins—a natal gift from its mother—the fruit fly embryo has sketched a floor plan for a body. It has marked the location of front door and back porch, calculated the depth of the cellar and the height of the roof. With the foundation completed and the framework begun, the embryo's next task will be to add detail to the pattern, subdividing the anterior-posterior and dorsal-ventral axes to define "a

---

*Although it seems backwards (or upside down), Dorsal actually activates genes specific to the ventral portion of the fly. Fly geneticists name genes based on what goes wrong when they're mutated, rather than on their function in the normal embryo. An embryo with a derelict *dorsal* gene is all back and no belly, hence the perverse name.

coordinate system that can be used to specify positions," Kirschner's invisible anatomy of compartments.

In the anterior-posterior direction, the interaction of the four products of the so-called maternal effect genes—Bicoid (front-to-back), Hunchback (front-to-back), Nanos (back-to-front), and Caudal (back-to-front)—defines a series of bands characterized by unique combinations of transcription factors. Within these crude compartments, those combinations trigger the expression of a second wave of embryonic genes, known collectively as the gap genes. New transcription factors encoded by the gap genes then diffuse and cross paths themselves, creating a series of even smaller compartments and recruiting the so-called pair-rule genes to staff them, each expressed in seven alternating stripes. Together, they partition the embryo into a total of 14 divisions, known as "parasegments," a temporary organization reminiscent of—but slightly out of register with—the familiar segmental pattern of the fly larva. Now, after a blissful period in which diffusion knew no bounds, the developmental process introduces a new complication: cells. Fingers of plasma membrane separate and surround the nuclei, putting a stop to this nonsense of using transcription factors as morphogens. It's time to break out the signaling proteins.

In the border country where one parasegment is to end and the next should begin, a pair of complementary signaling molecules initiates a quarrel that will establish the line dividing one from the other. At the posterior edge of each parasegment, a band just one cell wide marks its side of the boundary with the word "wingless." When the cells on the other side of the boundary—the anterior end of the next parasegment—hear their neighbors posturing, they reply with an admonition of their own: "Hedgehog!"* "Maybe you didn't hear us the first time," the first cells retort, "Wingless!" "You're the ones who can't hear," their antagonists counter. "Hedgehog! Take that!"

---

*So named because mutants that don't have this word in their vocabularies sport an unruly thatch of tiny spines.

Tit-for-tat, the two carry on like a pair of fishwives, locked in a self-perpetuating tape loop that demarcates a clear and stable boundary separating each parasegment from the next in the sequence.

The final step in the construction of the anterior-posterior "invisible anatomy" is the activation of genes charged with locking in the longitudinal pattern and specifying the anatomical features characteristic of each subdivision in the mature fly, from the mouth parts in the front of its head to the segments of the abdomen. Known as the "homeotic selector" or "Hox" genes, they encode gene regulatory proteins and are arranged along chromosome 3 of the fly in the same linear sequence as the structures they specify. Heading the lineup is *labial*, which defines the architecture of the head. A little farther along the chromosome, *Antennapedia* governs the placement of a pair of legs and a pair of wings, while *Ultrabithorax* defines the segment responsible for the balancing organs known as halteres. *Abdominal A* and *B* take charge of the abdomen. Because they determine the identity of each segment, mutations to Hox genes scramble the head-to-tail organization of the body plan. Mistakes in *Antennapedia*, for example, produce flies with antennae where they ought to have legs, while misexpression of *Antennapedia* in head segments yields a fly with legs sticking out of the top of its head.

The maternal effect genes, the gap genes, and the pair-rule genes are transient phenomena, but once the Hox genes are activated they will not be turned off. Regulatory proteins included among their targets maintain the status quo in each segment, locking the subset associated with that segment in the "on" position and mothballing the rest. As a result, the patterns invoked by these genes are set in stone, irrevocably committing cells that express them to specific fates, appropriate to their location along the length of the embryo.

The interactions of signaling molecules and the genes they manipulate partition the embryo into smaller compartments along the dorsal-ventral axis as well. In the nether regions farthest from the maternal signal that releases Dorsal from bondage, the silence is pierced by a childish voice chanting "Decapentaplegic. Decapentaplegic." This immense musical word is an embryonic antonym for Dorsal, a diffusible protein signal spreading downward to form a countergradient in the dorsal-to-ventral direction. Together, the complementary gradients of Dorsal and Decapentaplegic reinforce the definition of the top and bottom of the embryo much as the complementary gradients of Bicoid and Nanos reinforce the definition of head and tail.

Time out for a vocabulary lesson.

Decapentaplegic is your introduction to another clan of signaling proteins—actually, an offshoot of the extended family headed by a patriarch called "transforming growth factor-β" (TGF-β). Numbering about 20 members altogether, they are known collectively as "bone morphogenetic proteins," or BMPs, because of their felicitous effects on bone growth. All are actually twins (though not necessarily identical), a pair of polypeptide chains shaped "like an open hand with a pair of extended fingers." The two live and work together, the "fingers" of one cradled in the "heel" of the other. And all 20 take similar compound verbs, a complex of two receptor kinases targeting the amino acids serine or threonine (instead of tyrosine). The bone morphogenetic protein straddles these two receptor serine-threonine receptor kinases, prompting the one called "Type II receptor" to add a phosphate group to its Type I receptor partner. Once activated, the Type I receptor kinase then phosphorylates a relay protein called "Smad." Two Smads make a transcription factor; so the Smad decorated with phosphate by the receptor recruits a confederate—a different SMAD—and the two sashay into the nucleus, ready to translate the BMP message into changes in gene expression.

When Decapentaplegic and Dorsal are finished flipping genes on and off, the fly embryo has been partitioned into a series of compartments that run perpendicular to the compartments mapped out by the maternal effect genes and their successors. In the bottommost compartment, high concentrations of nuclear Dorsal have activated the genes *twist* and *snail*. Just a smidgen to the north, the concentration of Dorsal is no longer high enough to support *snail* expression; here only the Twist protein is produced. As Snail peters out and Dorsal declines, the *rhomboid* gene comes on. Finally, near the equator, Dorsal's whisper activates the *short gastrulation*, or "*sog*." For Decapentaplegic, running into the Sog compartment is like running into a brick wall—Sog hates the bone morphogenetic protein and silences it immediately. Cells able to hear Decapentaplegic give rise to the ectoderm of the fly's back. But where the Decapentaplegic signal has been censored by Sog, cells elect a different fate: they become the fly's nervous system. When the signal trading and gene tinkering are over, the embryo can stand back and admire the completed invisible anatomy of domains, each with a specific address defined by the intersection of anterior-posterior and dorsal-ventral gradients.

Meanwhile, subdivision of the dorsal-ventral axis has dovetailed with a dance in which the single layer of cells formed when the syncytium went cellular has shape shifted into three. Embryologists call this reel "gastrulation," from "gastro," meaning "gut." Initiated when the cells representing the underside of the embryo—the cells that do the *twist* and *snail* routine in response to ear-splitting levels of Dorsal—draw together and roll inward to form a tube, gastrulation masterminds the all-important goal of differentiating inside and outside. In the process it introduces groups of cells eager to make new friends and carry on conversations, relationships that will be the takeoff point for the genesis of the nervous system, the musculature, and the internal organs.

OUTSIDE IN

It swims! It tumbles! It chases its prey! While the barely multicellular *Volvox* dog-paddles between sunny spots, *Hydra*, a minute relative of the jellyfish living in the same pond, flips and glides in pursuit of the one-celled organisms it calls dinner. *Hydra* can swim circles around its algal comrade because its ancestors upgraded their locomotor apparatus, trading beating cilia for a new type of cell: muscle. Not that you'll find red meaty muscles bulging under its skin. *Hydra* doesn't have a skin; it's just a pair of epithelial sheets, rolled up into a tube and crowned with a fringe of tentacles, and its muscle power doesn't come from organized cohesive organs like our biceps but from primitive contractile cells, outfitted with elastic protein fibers able to contract and pump the animal along. These muscle-like cells are not even grouped together; they're distributed randomly throughout the animal, both the outer epithelium, or ectoderm, and the inner layer, known as the endoderm.

Somewhere between the ancestors of *Hydra* and its kin and those of the flatworms —the next branch on the phylogenetic tree—evo-

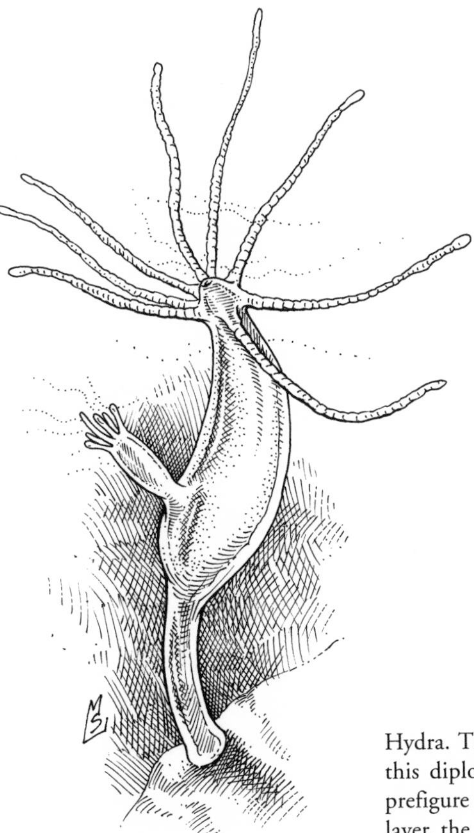

Hydra. The primitive contractile cells of this diploblastic (two-layered) organism prefigure the development of a third germ layer, the mesoderm, in larger animals.

lution took the muscle idea and expanded it into an entirely new family of tissues. This "mesoderm" (so named because it was sandwiched between the ectoderm and endoderm) gave rise not only to the musculature but also, as the metazoan lineage evolved, the digestive organs, the circulatory system, and the skeleton. An architectural innovation as significant as the post and lintel or the flying buttress, "the evolution of the mesoderm enabled greater mobility and larger bodies" and was a prelude to the evolution of such sophisticated features as a central body cavity, or coelem; a so-called through gut traversing the length of the organism; and appendages such as arms, legs, and wings.

The tripartite organization of larger metazoans is not easily appreciated in the adult, where it has been obscured by the intricacy of the mature internal anatomy but can be readily observed in the embryo—a fact that did not escape the notice of the nineteenth-century embryologists who spearheaded the acceptance of epigenesis. Indeed, the discovery of the three so-called germ layers was one of the strongest arguments in favor of progressive development, as each was found to give rise to a different set of tissues: the skin and nervous system from the ectoderm; muscles, bones, kidneys, heart, and blood from the mesoderm; and the lining of the gut, trachea, and lungs from the endoderm.

What's more, nearly a century before Hans Spemann's experiments with lens induction, Christian Pander, who discovered the germ layers, contemplated the possibility that the fates of the three layers might be interwoven, that "although already destined for different ends, all three influence each other collectively until each has reached an appropriate level." Today, we know that the migration of cells and the determination of their identities that occur during the formation of the mesoderm—the central element in the dance of gastrulation—are indeed the result of intimate conversations. The induction of the mesoderm, the first steps in the construction of the head, the migration of cells, and the partitioning of the ectoderm into cells that will form skin and cells that will form brain are all orchestrated by chemical signals exchanged between embryonic cells. Under the direction of these signals, "cells required for the formation of specific organs or body parts are physically brought together. Such juxtapositions of tissues, either transiently or permanently, facilitate inductive interactions that are critical for lineage specification and tissue patterning."

After fertilization the amphibian egg cleaves to form a cluster of smaller cells, weighted at one end (the vegetal pole) by the bulky yolk that will nourish the growing embryo. Near the other end, the animal pole, adhesive contacts pull cells apart to create a small cavity, known as the blastocoel, creating space for the cells about to begin their great migration. Then, at a point just shy of the equator, the surface of the embryo dimples. This indentation—embryologists call it the "blastopore"—widens into a smile, a parting of the lips that invites nearby cells to explore the darkness within. Those cells—beginning with the adventurous individuals poised at the blastopore's upper lip, dive eagerly into its maw, slide down its throat, slither along the roof of the blastocoel, and come to a stop at the far side. As they go, these adventurers pull the cells behind them down over the outer surface of the embryo like a stocking cap. In effect, the embryo swallows itself whole. The slipping and sliding transform the ball of cells into a hollowed-out plum, with the cells of the primordial endoderm lining the pit, those representing the immature ectoderm forming the skin, and, sandwiched between the two, the soft fruit of the future mesoderm, idling in anticipation of further instructions from the neighbors.

Contact with the embryonic optic vesicle could persuade cells undecided about their futures to devote themselves to forming the lens. Once they accepted the job of lens maker, however, cells lost the right to change their minds. Hans Spemann found that gastrulation was a time of equally momentous and irrevocable decisions. Cells started the dance of the germ layers open to suggestion—for example, a snippet of tissue transplanted from a region that would normally become a tadpole's nervous system to a region slated to become skin followed the example of its new neighbors and grew up to be skin instead of nerve:

> The first experiment consisted in exchanging a portion of presumptive epidermis and neural plate between two embryos of the same age, each being at the beginning of gastrulation. The grafts took so smoothly and development proceeded so normally that their margins left no trace. . . . From this it was obvious that, as we expected, the portions were interchangeable.

Spemann and a student, Hilde Mangold, discovered one exception, however. Nerve and skin spent most of gastrulation deliberating before finally deciding on a fate and committing to it, but the precocious cells that formed the upper lip of the blastopore already knew what they wanted to be before the process even began:

> It became apparent that a limited area, namely, the region of the upper and lateral blastopore lip did not conform. A portion of this kind, transplanted in an indifferent place in another embryo of the same age did not develop according to its new environment but rather persisted in the course previously entered upon and constrained its environment to follow it.

In one of the most celebrated experiments ever conducted by developmental biologists, Spemann and Mangold transplanted a piece of the dorsal blastopore lip taken from a newt embryo of a dark-colored species into the underside of an embryo of a light-colored species. But instead of switching gears and making belly skin (as ectoderm exposed to the optic vesicle responded by forming a lens), the blastopore transplant imposed its own agenda on its new neighbors. It ordered *them* to change direction, to put aside conventional ideas of becoming skin and take up the construction of a second backbone and spinal cord instead, setting in motion a sequence of events that culminated in the formation of a Siamese twin, joined

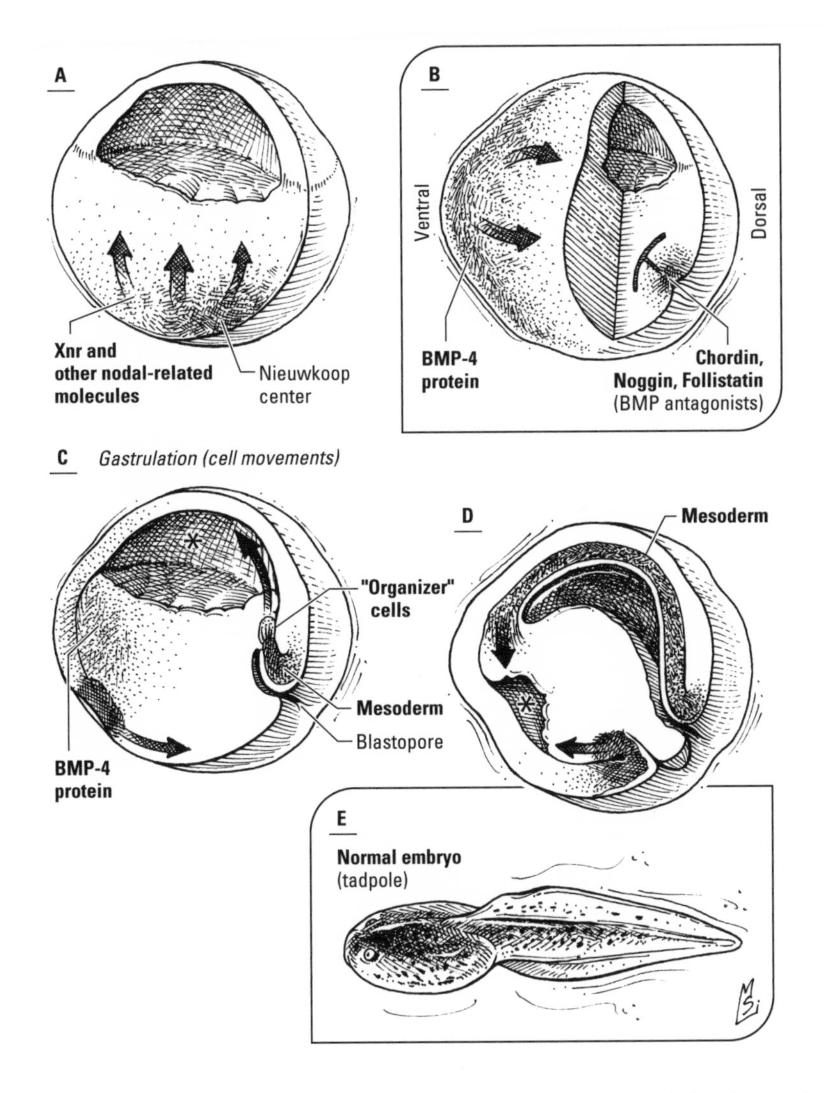

The induction of the mesoderm, the formation of the organizer, and the dance of gastrulation. *A*, Nodal-related proteins like Xnr, secreted by part of the primordial endoderm known as Nieuwkoop center, tell overlying cells to become mesoderm. Where concentrations of Xnr are highest—the dorsal lip of the blastopore—cells form a signaling center known as the organizer (*B*). The involution of the newly formed mesoderm, beginning with the cells of the organizer (*C, D*), segregates the three germ layers and establishes the dorsal-ventral and anterior-posterior axes, while the interaction of bone morphogenetic proteins, made by the ventral mesoderm, and BMP inhibitors, made by the organizer, differentiates the cells that will give rise to the nervous system from those that will give rise to the epidermis. *E*, the completed tadpole.

belly to belly with the host embryo. Because of their power to restructure the embryo, Spemann called the cells of the dorsal blastopore lip the "organizer."

A benevolent dictator, the organizer never resorted to pushing cells around but won them over with compelling arguments. Spemann's students Victor Twitty and M. C. Niu demonstrated that fluid drawn from a culture dish containing cells of the dorsal blastopore lip was as effective at duplicating embryos as the tissue itself, proof that the organizer's authority resided in secreted chemical signals, not direct contact. In fact, researchers learned, the organizer itself is the product of earlier inductive interactions. The amphibian egg, it turns out, is talking from the moment it's fertilized, conversations that not only lay the groundwork for the great migration of gastrulation but also determine the dorsal-ventral orientation of the embryo, the placement of the nervous system, and the boundaries of that all-important third germ layer, the mesoderm.

The frog embryo isn't spoon-fed all the essentials of a floor plan like the fruit fly. Still, it is not dropped clueless into the pond. Dad offers the infant one hint: "dorsal" is the side opposite the point where the sperm pierces the egg. And Mother provides a second, in the form of mRNA encoding the gene regulatory protein β-catenin, essential confederate of transcription factors currently idling in the cell nucleus. But like Dorsal, β-catenin will never see the inside of the nucleus unless it receives explicit permission, in this case the transmembrane protein Dishevelled, implanted in vesicles sandwiched between the egg cortex—the band of cells lining the perimeter of the egg—and the yolk. Sperm entry releases microtubules stacked just under the surface of the egg's plasma membrane like cordwood; down they slide, rotating the egg cortex so that the vesicles line up with the β-catenin protein on what fertilization has determined will be the dorsal side of the embryo. Now, Dishevelled can stabilize the transcription factor in this dorsal region, allowing β-catenin to accumulate to levels that can be heard by dorsalizing genes.

Meanwhile, under the influence of another transcription factor tucked into the egg by the mother frog, cells deep in the interior of the southern hemisphere—the primordial endoderm—establish radio station WXNR, broadcasting the same playlist around the clock: variations on the song "Xnr," short for "*X*enopus (the frog species most popular with biologists) *N*odal-*r*elated" proteins. Members of the TGF-β family of signals that includes the bone morphogenetic proteins, Nodal proteins can be heard on the air in all developing vertebrate embryos; frog, fish, bird, or mammal, they send the same message: make mesoderm.

It's easy to spot XNR's loyal listeners. They're the ones a little to the north, straddling the equator, particularly the cadre sitting on top of the blastopore, articulate types that entertain fond memories of Hans Spemann. Under the combined influence of stadium concert volume Xnrs and the dorsalizing influence of β-catenin, they not only agree to be mesoderm—specifically, dorsal mesoderm—but also volunteer for the all-important job of organizer. Their neighbors to either side appreciate XNR as well, although they don't have it playing at the same volume and they don't enjoy the simultaneous benefits of exposure to β-catenin. Still, listening to a Nodal protein is enough to convert them into lateral mesoderm, responsible for the heart, blood, blood vessels, and limbs (both naturally occurring and experimentally induced). On the side of the embryo opposite the organizer, however, XNR reception is so poor it's barely audible. Cells here listen carefully and take the mesoderm message to heart but interpret it their own way; they become ventral mesoderm.

As gastrulation gets under way, cells appointed to the mesoderm by Nodal signaling proteins begin to dive into the blastopore and slide along the roof of the egg, beginning with those that will form the head structures. Next come the cells responsible for the connective tissues—bone, muscle, and cartilage—of the back, followed by heart and kidney mesoderm. Bringing up the rear are the cells of the ventral mesoderm. Following in their wake, the cells of the future ectoderm wrap themselves around the surface of the embryo, swaddling the outer surface of the embryo in tissue that will form skin and nervous system.

Every cell in that ectoderm wants to be a neuron. And theoretically it could, for that's the default condition, the path an ectodermal cell is slated to follow at this point unless it's told otherwise. Unfortunately, some must take the less glamorous job of skin. The ventral mesoderm, assigned the thankless job of posting the bad news, heaves buckets of the bone morphogenetic proteins BMP4 and BMP2, synonyms for "ventral" and "skin," at the ectoderm. But if everyone changes course and follows these instructions, the embryo will be all belly skin and no brain. It's up to the organizer to temper the situation, issuing antagonist proteins that intercept the BMPs and negate them. "Chordin!" it yells. Or "Noggin!" Or "follistatin!" Cells exposed to the highest concentration of these molecules ignore the BMP message and follow the default program, remaining dorsal and becoming neural. Cells farther from the organizer, exposed to plenty of BMP signals but little of the antagonists, form epidermis. Collectively, the intersecting gradients of BMPs and organizer antagonists partition the ectoderm, carving out one territory to be the nervous system and the other to be the skin that protects it and other underlying structures from the elements.

Creatures like birds and mammals also turn outside in during embryonic development but have altered the vertebrate developmental program exemplified by the frog to accommodate changes to the basic body plan as well as the evolution of extra-embryonic tissues of the placenta. In the chick, for example, the embryo, cramped into the limited space between the hard eggshell and the massive yolk, begins as a disc of cells, the blastoderm, instead of a ball. Selected cells of the mammalian embryo secede early in development to form the placental tissues; the embryo itself is cup or disc shaped. Cells of both migrate into the interior of the embryo through a groove known as the primitive streak, rather than a blastopore—but both still feature signaling centers that control the formation and patterning of the germ layers.

What's more, put your ear to a hen's egg or a mother mouse's belly and you'll hear familiar commands directing traffic. "Step back, please," calls β-catenin, and you know that this will be the dorsal side of a chick embryo. Nodal proteins still induce the mesoderm. Where high levels of Nodal and β-catenin coincide, an organizing center, known as Henson's node in birds and simply "the node" in mammals, forms and secretes signals antagonistic to bone morphogenetic proteins—a contest that still figures prominently in the decision to become neural ectoderm or epidermal ectoderm.

To the chorus of signaling molecules belting out mesoderm and migration, these embryos have added another: the voice of FGF, harmonizing with Nodal proteins, BMPs, and BMP antagonists. In the early embryo, FGF joins the call for the induction of the mesoderm. Later in development, the meaning of the FGF signal changes, from "make mesoderm" to "make nervous system." Secreted by precursors of cells that form Henson's node—the avian equivalent of the amphibian organizer—FGF "primes" cells of the chick embryo to respond more enthusiastically to the proneural, anti-BMP messages that will follow.

According to Lewis Wolpert, "The most important event of your life is not birth or marriage, but gastrulation"—a milestone you share with frogs and fish, the chick and the mouse. The prelude to this dance is the appointment of cells to the mesoderm; its climax, the sorting and shifting of these cells and their compatriots to form the three germ layers. Along the way, the order of migration and the interactions between signals, particularly the bone morphogenetic proteins and their antagonists, have also patterned the anterior-posterior and dorsal-ventral axes and sketched the location of the nervous system. Choreographed by signaling molecules conserved across species, the balletic movements of gastrulation, coupled with the events of earlier development, culminate in the solution of problems that confront all vertebrates and, indeed, all animals: the establishment of polarity, the isolation of an internal milieu, and the generation of an invisible grid of compartments in preparation for the next stage of development.

AN ARM AND A LEG (AND A BACK AND A BRAIN)

The foundation has been poured, the walls erected, the roof shingled. But it will take more than two-by-fours to make this skeleton of a house a home. Windows and doors must be hung. The future kitchen, bathrooms, and laundry room must be plumbed. The new owners will surely want to move around, cook, read, supervise homework, or find their kids after dark, so an electrician is in order. Someone had better grade that hillock of dirt out front, and someone else had better turn that floodplain of mud into a driveway. And that's not even counting all of the amenities—carpets, appliances, fixtures, tile—still to be installed.

The embryo's construction project has also made considerable progress. Signaling pathways and teams of transcription factors have decided which end will lead and which will follow as well as which side will face the sky and which the ground, assigned cells to germ layers, rearranged cells to seal off the interior of the organism from the outside world and facilitate the further exchange of chemical signals, and divided the embryo into a system of compartments, each characterized by a unique pattern of gene expression. Now these compartments can begin to interact to refine positional information, shape organs, put the finishing touches on the body, and drive the final differentiation of cells and tissues. "The initial structure is the basis for further subdividing the organism into smaller and smaller domains," explains Marc Kirschner. "Once an embryo has mapped out the initial spatial pattern of compartments, cells within the domains can further differentiate in a variety of ways. For example, those domains can interact with neighboring domains. And as a result of those interactions, you can end up with cells that are now different because they were close to a different domain."

In a typical vertebrate embryo, for example, the molecular

machinations of early development have defined, along the longitudinal axis, the rough outline of the torso. The embryo has a top and a bottom; clearly, the future limbs go somewhere in the middle. But while there's a space to put an arm or a leg, it's impossible to make out even the slightest bulge of an emerging limb bud, much less a fully jointed limb complete from shoulder to fingers or hip to toes. The mesoderm is just a layer of cells; actual bones and muscles are still in the planning stages. And the cellular territory labeled "neural ectoderm" by FGF and BMP antagonists like Chordin and Noggin doesn't constitute a functioning nervous system any more than a stack of bricks and a pail of mortar constitute a finished wall.

The developing embryo meets the challenge of later embryonic development the same way it met earlier challenges: by talking itself along toward maturity, linking signals to transcription factors that continue the task of "compartmentalizing" the genome, as Kirschner calls it. Of course, more compartments—and more interactions between compartments—demand precise, effective communication. Some tasks will require new words and novel variations on the basic syntax of cellular communication. But familiar words and tried-and-true mechanisms, such as the morphogen gradient, can also be used in new ways; indeed, the parsimonious nature of evolution encourages such recycling. "Setting up a graded positional cue is unlikely to be simple," according to Miriam Osterfield, Marc Kirschner, and John Flanagan, in a 2003 review published in the journal *Cell.* Given the time and effort needed to elaborate the morphogens themselves, the components of their signal transduction pathways, and mechanisms to control signal secretion, deposition, and action, "it is not surprising that evolution, having set up a successful gradient system . . . would exploit it for multiple purposes." Hedgehog, the signaling protein that meant "posterior" in the fruit fly during the definition of the parasegment boundary, is a good example. Swap an amino acid here and there to put a vertebrate stamp on it, give it a new definition: "motor neuron" and you can use it to turn a sheet of ectoderm into the beginning of a central nervous system.

"What we actually know about Hedgehog is provocative, even if it's not definitive," says molecular biologist Philip Beachy. "To begin with, the signal is made in an interesting way." Beachy explains that Hedgehog (actually, Hedgehog*s*, for vertebrates like us make three different varieties of Hedgehog protein: Desert hedgehog, Indian hedgehog, and Sonic hedgehog), like many other proteins, begins life as part of a larger precursor. But while the others depend on a separate protein-trimming enzyme, or protease, to liberate them, the Hedgehog precursor does the job itself. Even more curious, the knife it uses bears an uncanny resemblance to a type of protein domain more typical of bacteria. Known as an "intein," it "carries out a splicing operation. The intein excises itself, then rejoins the two bits of the 'host' protein," says Beachy. Then again, Hedgehog seems to collect heirlooms: another segment of the protein can be superimposed on a common bacterial cell wall enzyme. Hedgehog, in other words, may be a chimera, two prokaryotic enzymes fused into a single eukaryotic signaling protein. "What sort of organism this happened in or what pathway it took from primitive unicellular organism to patterned multicellularity, we don't know," Beachy notes. What's important, he adds, is that the journey culminated in a protein with a topography that "is useful for binding."

Then there's Hedgehog's choice of fashion accessories. It's not unusual for eukaryotic proteins to don bits of frippery during the final stages of synthesis, typically sugar groups thought to expedite folding or discourage assaults by proteases. Hedgehog, however, believes that sugars are *so* last millennium. Its jewelry is made of lipids: the fatty acid palmitate and a molecule of cholesterol. Now decorating yourself with lipid might be creative, but that's just not the way it's done in the Fraternal Brotherhood of Secreted Proteins. Biological membranes, you'll remember, are principally lipid themselves. A lipid-decorated molecule, therefore, is more likely to be trapped in the membrane than to be secreted. With the spare tire it's sporting, Hedgehog ought to be about as mobile as a sumo wrestler. Instead, it's gotten around the problem with its own personal taxi service. A receptor-like protein, Dispatched, binds and escorts Hedgehog out of the cell, then inches along a path of sticky, sugar-coated proteins embedded in the extracellular matrix to place the signal exactly where it's needed. "The regulation of Hedgehog movement is so important it requires a dedicated component just to release the signal," Beachy says. "You can think of cholesterol as the 'handle' that directs Hedgehog to this movement machinery. It doesn't so much restrict [Hedgehog's] motion as facilitate its regulation."

Finally, Hedgehog's grammar would have your high school English teacher reaching for her red pencil. Every Hedgehog sentence contains a double negative: the signal itself is a synonym for "no," and so is its receptor, Patched. Blame the arrangement on the chatterbox heading the signal transduction relay, a second transmembrane protein, Smoothened, which shadows the Patched receptor much as a remora shadows a shark. Smoothened loves to talk; left to its own devices, it would have the pathway's target genes on every hour of every day. And since that would encourage not only some odd developmental decisions but uncontrolled cell division as well, Smoothened needs to be hushed unless there's a legitimate reason for it to be active, a job that falls to Patched. In the absence of Hedgehog, the receptor whispers "No talking!" and enzymes farther down the pathway lop a bit off the transcription factor known as "Ci" (in flies, for "Cubitus interruptus") or "Gli-1, -2, or -3" (in vertebrates, for "glioblastoma," a kind of brain cancer), turning it into a gene repressor rather than a gene activator. When Hedgehog says "no" to Patched, Smoothened can tell the enzymes to leave the transcription factor alone; in its intact form, it can now turn genes on.

The double negative is there for a reason: it forces cells to stop, look, and listen before making a life-or-death decision. As Beachy points out, even after Hedgehog has shut Patched down, gene expression doesn't flip from off to on immediately. "The repressor [that's already present] must be degraded, over a period of hours, before you have full activation," he says; as a consequence, "you have a built-in time delay in the circuit." He speculates that this lag between pathway activation and gene activation may ensure a critical level of precision. "In these pathways you have a longer timeframe. Precision is more important than speed. You don't want to be set up on a hair-trigger; you want to take the time to get it right, because it has such critical implications for the development and life of the organism. Make a mistake and you could end up with the wrong pattern."

If there's one place where a vertebrate embryo certainly doesn't want to end up with the wrong pattern, it's in the developing central nervous system.

Growing up is never simple. But perhaps no tissue faces a more daunting challenge than the nervous system. Beginning only with the dorsal ectoderm rescued from a life as skin by the antagonists of the organizer, the vertebrate embryo must shape a brain at the head end and run a spinal cord down the back. It must tell some cells to be informatics specialists—neurons—and others how to become the glial cells that insulate nerve fibers and assist the neurons in their work. What's more, the decision making doesn't stop here. A cell that has elected to be a neuron must still decide what sort of neuron—perhaps a squat, bushy Purkinje cell in the cerebellum, a willowy pyramidal cell in the cerebral cortex, or an interneuron shuttling messages over short distances—and it must be grouped with similar neurons to form collectives. Finally, the peculiar demands of information processing have made neurons very choosy about who they'll talk to. Each must thread its axon, the fiber it will use to talk, precisely through the amorphous darkness to its appointed partner—even if that partner is located on the other side of the brain.

Thanks to the organizer's zealous propagandizing, cells of the dorsal ectoderm have bought into the neural agenda. Now they will grow tall, into a layer of long, thin cells known as the neural plate. Pushed by the epidermis on either side, the neural plate hunches up and begins to fold in half along the midline like a book. The longer the epidermis shoves, the higher and closer the edges of the plate grow, until they meet and merge, overrun by future skin. Fluid trapped in the anterior end strains at the walls of the tube and causes them to balloon outward in three places, forming the so-called primary brain vesicles. As neural development proceeds, two of these vesicles will be squeezed in the middle to create a total of five bulges at the anterior end of the neural tube corresponding to the five major divisions of the brain: telencephalon, diencephalon, mesencephalon, metencephalon, and medulla.

The posterior portion of the neural tube will become the spinal cord. This part of the nervous system is home to the motor neurons that command the muscles; the way station where incoming axons of the sensory neurons, transmitting information from skin receptors mediating touch and pain, contact neurons charged with relaying the data to sensory processing centers in the brain; and the work space of a multitude of interneurons, cells that ferry information from neuron to neuron rather than from sense organ to neuron or neuron to muscle. In the mature spinal cord, these neuronal elements form a central core, the gray matter, embedded in a cocoon of insulated fibers—the so-called white matter—traversing the cord's longitudinal axis. Viewed in cross section, the gray matter stretches out in a butterfly-shaped formation splayed across the midline: motor neurons in the so-called ventral horn (the lower wings of the butterfly), incoming terminals and sensory relay neurons in the dorsal horn (the upper wings), interneurons concentrated in the deeper layers near the center. Here in the embryo, however, the future spinal cord languishes in shapeless anticipation, while the ectoderm overhead and the mesoderm below—specifically, the notochord, a temporary stand-in for the backbone while it's under construction—contest for the hearts and minds of immature cells.

If you detect something familiar about the notochord, you're not imagining things; the cells of this structure are descendents of the organizer, all grown up and done with baby talk. Now their vocabulary consists of the words "Sonic hedgehog," which they chant ceaselessly at the base of the neural tube. Cells that listen to Sonic hedgehog and take its message to heart become the floor plate and begin to chant "Sonic hedgehog" themselves. Their signal diffuses out of the floor plate and generates a gradient, pointing in a ventral-to-dorsal direction.

At the same time, the epidermis above the neural tube is also talking. It spews a cocktail of signaling molecules belonging to the TGF-β "superfamily," in particular the bone morphogenetic proteins BMP4 and BMP7. These signals coax cells at the top of the neural tube into becoming the roof plate and setting up their own signaling center. Roof plate TGF-β signals diffuse out of the plate region and create a second gradient, pointing in the dorsal-to-ventral direction, a mirror image of the gradient formed by Sonic hedgehog.

Depending on where they are located, cells of the neural tube are exposed to varying amounts of Sonic hedgehog and TGF-β-type signals and base their decisions on what to do next accordingly. Cells near the roof plate, where TGF-β signaling predominates, turn on transcription factors that start them down the road to becoming interneurons. Those just north of the floor plate pay more attention to Sonic hedgehog. They activate transcription factors able to help them become motor neurons eager to learn which muscle group they are to innervate—perhaps, in a hypothetical world, the muscles of your third arm.

Cells can be neither motor neurons nor interneurons, however, unless they've committed irrevocably to the neuronal program. Membership in the neuroectoderm is a job interview, not an employment contract; to secure a permanent position in the company of neurons, a cell must first face off against any number of equally qualified neighbors. "Contingency plans are part and parcel of development," writes developmental biologist Eric Lai. "Often, more cells than necessary have the opportunity to become a specialized cell type, a scheme that allows for backups." But when all applicants look the same and express the same genes, how is a little embryo to decide who ought to get the job? Fortunately, cells can resolve this issue themselves, calling up a novel signaling mechanism evolved specifically for conducting one-on-one conversations and consisting of just two words: "Notch" and "Delta."

"Unlike other signaling pathways, which are activated by factors made elsewhere in the embryo, Notch signaling takes place right between cells," says molecular biologist Gerry Weinmaster, an intimacy necessitated by Delta's agoraphobia. She explains that while hormones sail the bloodstream, growth factors diffuse, and Hedgehog takes a taxi, the Delta signal stays rooted firmly in the membrane of the cell that made it, daring to expose only its head and shoulders. Notch, the receptor, is also embedded in the membrane. In order for Notch to bind Delta, therefore, the cell bearing the receptor and the cell sending the signal must be close enough to touch.

It shouldn't come as a surprise, then, that both cells participate in the signaling process. The Notch receptor doubles as this pathway's transducer, but before it can put on its transduction hat, it has to escape from the membrane. That's where the Delta-bearing cell steps in to help. As soon as Notch binds Delta, the receptor stretches its neck out so that a nearby protease can behead it; the signaling cell then swallows both the signal and binding site fragment of the receptor (to which it's still attached) in one gulp. With that mess out of the way, a second protease can swoop in and amputate Notch's tail. This disembodied free end is the pathway's only intracellular signaling protein, and it hikes itself off to the nucleus to negotiate the exchange of transcription factors regulating so-called proneural genes. Whispers are followed by the muffled sound of the genes being wrapped for storage. No use pining—this cell's application to be a neuron has just been denied.

Notch and Delta's best-known accomplishment is the sensory bristle of the fruit fly, the arthropod equivalent of the touch receptors in our skin. Each bristle is actually a cluster of four cells—a neuron, the sheath cell surrounding it, a socket cell, and a shaft cell,

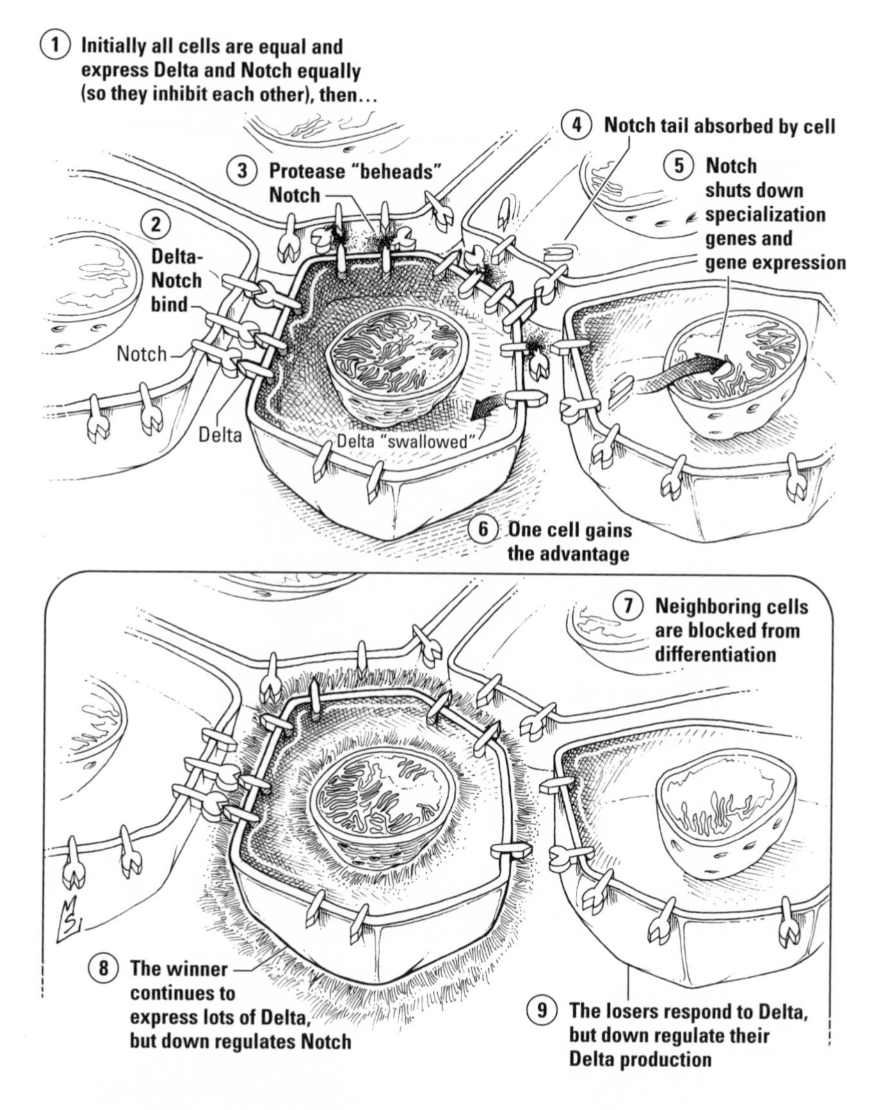

Lateral inhibition mediated by Notch-Delta signaling. Cells vying to become neurons begin as equals. At this initial stage, all express the membrane-bound signal, Delta, and its receptor, Notch; as a consequence, each inhibits its neighbors but is also inhibited by them. A lucky break gives one cell an advantage; as a result, it can suppress the so-called proneural genes of those in the immediate vicinity even more effectively, while downplaying its own sensitivity to the Delta they express, weakening any attempt to retaliate.

all descendents of a single sensory mother cell, winner of a competition between two to three dozen identical prospects. At the start of this contest, all the cells of this "equivalence group" express similar levels of both Delta and Notch; each, as a consequence, blocks the differentiation of its neighbors but is blocked by them in return. Development is locked in a stalemate—until one cell loses count and finds itself with a few extra Delta molecules, allowing it to talk just a little louder than those around it. This cell can now suppress its neighbors more effectively—including their expression of Delta—and tone down its own Notch receptors in the process, so it can evade their retaliation. As a consequence, the noisy cell gains the upper hand, while those surrounding it sink lower and lower into repression. Their proneural genes are turned off, and the loudmouth becomes the mother cell.

Cells in the neural ectoderm rely on a similar war of words to determine which will advance to the next step in the quest to become a neuron and which will change their minds. All express Delta and Notch initially. Then chance makes some stand out from the crowd, and feedback amplifies this small difference into a mighty advantage for the individualists. Accidents that shush Notch signaling—like a scientist inserting a garbled version of the Delta signal or engineering a mutation in the Notch receptor—rig the competition and pack the developing nervous system with superfluous neurons.

Marc Kirschner likens this cellular version of *Survivor*, called "lateral inhibition," to a power struggle in a primitive society. "It's like a group of people who are all the same, all equal. Then one of them becomes stronger than all the other ones, so they all become subservient and he gets more power and he can enforce his superiority. Finally, you get a king and you have a whole hierarchy of people because it's self-reinforcing. Two kings are unstable—they'll fight it out until there's only one. It always works out that you only have one."

Notch and Delta not only choose the precursors of the motor neurons, they help lay the groundwork for the construction of the muscles themselves.

Flies are divided into segments, vertebrate mesoderm, into somites. As the neural tube rolls up, blocks of the so-called paraxial mesoderm on either side break off at periodic intervals, beginning at the neck. These transient structures, the somites, are the starting point for the vertebrae; the ribs; the muscles of the back, trunk, arms, and legs; and the connective tissue, or dermis, of the back skin.

You can set your watch by somite formation. In a chick embryo, for example, the mesoderm spits out a new pair of somites every 90 minutes, like a skee-ball game at a carnival spitting out tickets at the end of each round. Open the back of this clock and you'll find Notch pulsing at the center of the mechanism. Each oscillation begins with a wave of Delta-inspired Notch activation that sweeps through the mesoderm and unleashes a battery of gene repressor proteins. One target of these regulators is the *notch* gene itself, setting up a negative feedback loop that suppresses Notch signaling. But the repressors themselves are inherently unstable. They collapse and decay, turning off the feedback mechanism and resetting the clock in preparation for the next round of signaling.

Notch oscillations are translated into a spatial pattern by another familiar signal, the growth factor FGF. Secreted by mesodermal cells as they march through the node during gastrulation, FGF keeps cells young. At the rear of the embryo, where concentrations of FGF are highest and its voice is the loudest, cells hear that they're too immature to assume the responsibilities of being a somite. But as they are pushed forward by newcomers, FGF levels decrease and the volume drops. At the leading edge of the FGF gradient, known as the "determination front," the message becomes too faint to influence the cells straddling the front. Finally ready to grow up, they stop listening to FGF and start listening for the clock; at its next tick, they break off into a somite.

Cells stationed along the lateral border of the somite—soon to be known as the myotome—are then coaxed into becoming muscle precursors by signals broadcast from the nearby lateral mesoderm and the overlying epidermis, in particular the Wnt proteins. Counterparts of the fruit fly parasegment boundary signal Wingless (the name Wnt comes from *w*ingless + *Int*-1, the first Wnt to be discovered), Wnt signals have much in common with Wingless's old partner, Hedgehog. To begin with, Wnt proteins are embellished with a fatty acid, palmitate. Similarly, a Wnt sentence also contains a double negative, with the signal saying no to a nay-saying receptor; the contested transcription factor in this case is β-catenin, the protein we met spelling out "dorsal" in the amphibian embryo. And once again, the reason transcription must be so carefully regulated is that β-catenin not only masterminds gene choices important in differentiation but also is a powerful stimulus for cell division. "If β-catenin was constitutively active," says developmental biologist Roel Nusse, cells would divide uncontrollably. So in the absence of Wnt, β-catenin is surrounded by a trio of bullies, the Degradation Complex, that attack it and turn it over to a protease before it ever gets near the nucleus. Wnt puts an end to this bloodshed. Its receptor, Frizzled, activates the intracellular signaling protein Dishevelled, and Dishevelled disperses the Degradation Complex. β-Catenin accumulates and "takes the Wnt signal to the nucleus," says Nusse.

The cells designated muscle precursors by Wnt signals migrate into the developing limb and activate genes for the transcription factors MyoD and Myf-5. So-called master switches, these proteins mark an irrevocable decision to become muscle and initiate a cascade of transcription factor activation that culminates in the expression of muscle-specific proteins. Encouraged to grow up by transcription factors and running out of the FGF they need to divide, committed myoblasts relinquish their childish interest in proliferation, adhere to, and then fuse with each other. Bundled into muscle fibers able to flex and contract, they are finally ready to move an arm or a leg.

In the laboratory, scientists can induce an arm, a leg, or a wing with a plastic bead and a little FGF, but of course "this is not the way a limb is normally induced," Cliff Tabin points out. Tabin admits, in fact, that it's not yet clear exactly how limbs get started in the living embryo—by the time "fibroblast growth factor" is first heard, Hox genes have already marked the boundaries of the building lot where each limb is to be constructed. What is clear is that building doesn't actually get under way until FGF is present and that those directing the workforce must take care to choose the correct suffix every time they say "FGF." We vertebrates actually have 22 variants of the word "fibroblast growth factor" in our signaling vocabularies; like the proverbial 100 words for snow in the Eskimo language, each has its own nuance. The variant that sets a limb bud growing is "fibroblast growth factor eight" (FGF8). Produced by cells adjacent to the somites, it tells the lateral mesoderm at the site chosen for the future limb bud to start making FGF10. FGF10 in turn talks to the overlying ectoderm, and it has two things to say: "Make your own FGF8 already!" and "Anyone interested in being part of the apical ectodermal ridge raise your hand." Without FGF10, the ridge fails to form; without the ridge, the mesoderm buds, but the bud never grows into a limb.

The FGF8 made by the ectoderm now takes over responsibility for maintaining the expression of FGF10 in the lateral mesoderm. In response, the mesoderm continues to promote the production of FGF8, "setting up a feedback loop," says Tabin. And that, he explains, is why an outside source of FGF—FGF8 *or* FGF10 (and FGF1, 2, and 4 as well)—can induce an ectopic limb. "When you implant an FGF bead, it induces FGF10 in the flank as if another apical ectodermal ridge was there. You're coming in the middle of that feedback loop."

Once the limb bud is established, the interaction of FGFs and other signals shape the limb as it grows, directing the formation of a cartilage template according to a pattern that is more or less the same in all four-legged animals. Gradually replaced by bone, this proximal-to-distal pattern consists of three elements: that closest to the body gives rise to the humerus, the large bone of the upper arm, and the femur of the thigh; the middle element generates the bones of the forearm (the radius and ulna) and leg (the tibia and fibula); and the third is responsible for the bones of the wrist, ankle, and digits. In addition, the limb, being a three-dimensional structure, must be patterned in two other directions as well. Just look at your hand. Along the anterior-to-posterior axis, patterning mechanisms have dictated the order of the fingers—thumb pointing toward the middle, little finger on the outside—while in the dorsal-ventral direction, they have flattened the hand and distinguished the back from the palm.

Just as the proximal-distal axis is specified by fibroblast growth factors, the anterior-posterior and dorsal-ventral patterns are spelled out for the growing limb by distinct signals. Back-to-palm patterning is the province of Wnts, specifically Wnt 7A, produced by cells in the ectoderm just above the apical ectodermal ridge. And the order of the digits is the responsibility of Sonic hedgehog, spoken by a cluster of cells nestled in the fold where the bud leaves the body wall. Recruited by FGF8 from the apical ectodermal ridge, this Zone of Proliferating Activity, or "ZPA," sets up its own reciprocal relationship, similar to the FGF10–FGF8 feedback loop that keeps the limb growing out from the shoulder, to keep the words flowing. "Speak to me," Sonic hedgehog from the ZPA says to the cells of the apical ectodermal ridge. "FGF4!"—code for "Play it again!"—they respond, and the cells in the ZPA repeat: "Sonic hedgehog!" As the arm elongates, the ZPA rolls along behind, muttering "Sonic hedgehog," and the benefit to you, oblivious one, is a correctly formed hand ready to grasp, pinch, hold, and throw.

Thirty years ago, Lewis Wolpert, who conceived the French Flag Model to explain how a morphogen gradient could dictate a cell's fate, also proposed a model to explain how chemical signals might generate the proximal-distal pattern of the developing limb. Wolpert

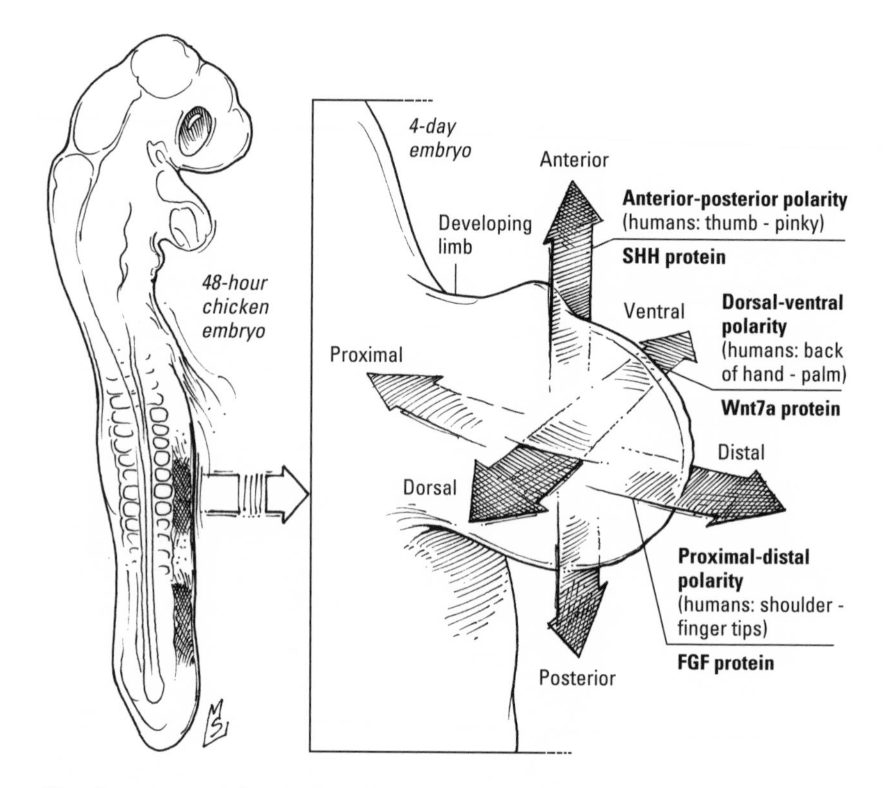

Signaling centers in the vertebrate limb. As the limb bud grows out from the body (a chick embryo illustrates the principle here), FGFs made by a signaling center in the apical ectodermal ridge establish the proximal-distal (shoulder to fingers) pattern; Wnt7A, made by the region to the north of the ridge, patterns the dorsal-ventral (back of hand to palm) direction; and Sonic hedgehog, made by cells in the so-called zone of proliferating activity, patterns the limb in the anterior-posterior (thumb to little finger) direction.

and others knew that removing the apical ectodermal ridge led to a truncated limb and that the extent of the deformity was a function of when the signaling center was amputated: removal early in development produced an arm that stopped at the end of the humerus, while postponing the surgery until a little later in development allowed the arm enough time to grow as far as the wrist. Wolpert proposed, therefore, that a cell's identity was a function of how long it kept up with the apical ectodermal ridge as the limb grew outward from the body. As long as a cell stayed in the ridge's shadow, it remained immature. But once it was pushed out of this "progress zone" by newly minted cells nearest the ridge, it stopped, checked the clock, and decided what to do next based on that information. For example, the first cells forced out of the progress zone formed the upper part of the limb; those that dropped out next, the middle element; and those that stayed the longest, the fingers and toes. "According to the model, cells got no instruction from the outside. Initially all the cells think they're supposed to be shoulder. As long as the apical ectodermal ridge is present, cells are constantly changing their minds: 'No we're not shoulder. We're an upper arm. No, we're not an upper arm. We're an elbow.' But if they get out of range of the apical ectodermal ridge, they can't change their minds any longer. That explains what happens if you remove the apical ectodermal ridge today and don't get an elbow tomorrow—the arm had specified up to the elbow," as Cliff Tabin explains it.

Tabin, however, is convinced the model is wrong. Label cells in the so-called progress zone before you remove the ridge, he notes, and you should find the marked cells afterward in the abbreviated humerus or femur, where the amputation had stopped them in their tracks. Instead, the cells disappear. Where do they go? That's simple, Tabin says. They kill themselves.

TWO STEPS FORWARD, ONE STEP BACK

To win at the survival game, animal species like humans, elephants, dolphins, and chimpanzees have evolved a "good mother" strategy. They reproduce one baby at a time, and they invest a great deal of time and energy in each infant. Thanks to this extra parental attention, the disadvantages of a slow rate of reproduction are offset by a better-than-average chance the infant will survive to adulthood. Insects and fish, frogs and clams, on the other hand, have adopted a reproductive strategy that might be described as success through excess. Instead of putting all their eggs in one basket, they spawn basketfuls of eggs. A female bullfrog, for example, can lay up to 20,000 eggs every time she reproduces—enough tadpoles, should they all hatch, to overrun the entire pond.

Plan for contingencies, and you're asking for clutter. If you don't believe it, take a look in your garage—or any mammalian embryo. Overstocking equivalence groups and padding primordial tissues with extra progenitors is a success-through-excess strategy that ensures the embryo won't come up short in the middle of constructing a heart, brain, or skeleton. But space and nutrients are just as limited in the embryonic environment as they are in the average pond. Once a building project is completed, the embryo must clean house or it will be overrun by cells that no longer serve any useful purpose.

When my daughters are ordered to clean their rooms unless they want their mother-chauffeur to go on strike, they push books and papers under the bed and abandon their laundry at the top of the stairs, but cells could never afford to be this careless. A wholesale massacre would create a mess, as the dead and dying left the area awash in a chemical spill that could prove equally lethal to the very cells the embryo just nursed to maturity and wants to keep. That's why every cell's natural inclination to kill itself at the drop of a hat—and do a neat job of it—comes in handy.

The locusts yellow first, within days of releasing their chocolate-colored pods. Then the maples and oaks, scarlet and gold. Finally, the ornamental pears turn, first orange, then crimson, then claret.

The trees prepare for this drama weeks before the first leaf changes, anticipating days too dark and short for photosynthesis and the loss of water, sequestered as snow. Sugars and proteins are drained from the branches and stored safely in the roots or trunk. Hormones that urged growth and flowering ebb. The junctions between stem and branch crack; finally the leaves simply fall away.

Leaves attacked by insects or disease, on the other hand, die under duress, spotted and curled. When the cells that make up the human body die a violent death, they don't make a pretty picture either. Damage to the cell's protective plasma membrane allows the surrounding fluid to leak in. Internal structures collapse. The cell swells and ruptures, dousing innocent bystanders with a cocktail of enzymes and waste products. Attracted by the commotion and the mess, white blood cells swarm to the scene of the accident, where they scavenge what's left of the victim—and release powerful chemicals that may kill the neighbors as well.

But not all cells die this miserably. In 1972, Scottish pathologists John Kerr, Andrew Wylie, and Alistair Currie published electron micrographs of a different type of cell death, a death that was quiet, orderly, even dignified. A cell that died this way withered, rather than exploded. It shrunk and its nucleus split, allowing the DNA to trickle out and clot. Bubbles erupted on the surface, broke off, and were swallowed by neighboring cells. Before all but its nearest neighbors were even aware of the death, the cell had vanished.

The process reminded Kerr of the falling of autumn leaves, and a colleague suggested that he call it "apoptosis," from the Greek *apo*, meaning "away from," and *ptosis*, "to fall." In contrast to the chaos of accidental death, apoptosis appeared to be "an active, inherently controlled phenomenon," a choice, not a consequence—in other words, a suicide.

Apoptosis offers cells that have outlived their usefulness, that exceed population quotas, or that have suffered grievous injuries a way to excuse themselves gracefully without further inconveniencing or endangering their neighbors. In the mature organism it provides a counterbalance to proliferation, keeping cell numbers at a steady-state level. During development it whittles overstocked compartments down to size, cuts away superfluous cells to shape organs and body parts, thins tissues so that cells selected for further differentiation have the space they need, and matches the number of cells to the availability of precious resources.

The suicide weapons of unwanted cells are a family of killer proteases. Known collectively as caspases (for *c*ysteine-containing, *as*-

*p*artate-targeted prote*ases*), they activate nucleases that degrade DNA, shred cytoskeletal proteins, initiate the collapse of the nuclear and plasma membranes, and trigger the activation and release of signals that encourage neighboring cells to scavenge the remains of the dead one. Too dangerous for a cell to leave lying around the house, caspases are normally bundled up as part of a larger precursor. Molecular time bombs, these packages drift, innocent and unnoticed, among the other cellular proteins, waiting for the opportunity to wreak havoc. "The cell is a society of proteins that interact with each other—they talk, send messages, go to meetings," according to molecular biologist Yuri Lazebnik. "Caspases are members of the society too, and all the time the cell is alive, they appear to be like everyone else. They smile at children, talk to the neighbors. But then they receive a signal of some nature and they become cold-blooded killers. They kill the neighbors, they kill the kids, they blow up the whole society, and then they die themselves."

Despite the precautions, eukaryotic cells were playing with fire, collecting lethal enzymes like caspases when they had black-hearted bacterial symbionts as permanent houseguests. Oxygen-loving prokaryotes may have withered away to mitochondria without a whimper, but one of their respiratory enzymes seems to have resented being put to work as slave labor. Chained to its oar and seething, cytochrome c bides its time and dreams of driving its host to suicide. As long as it remains sealed in the mitochondrion, it can do no more than stew in its grudges. Should an unlucky accident break down the prison walls, however, cytochrome c wastes no time turning its fantasies into reality. Pair bonding with a seedy character named "Apaf-1" (for *a*poptotic *p*rotease *a*ctivating *f*actor), it erects a factory dedicated to the production of active caspases, an act of revenge that guarantees the demise of the enzyme's unfortunate landlord.

The integrity of the mitochondrial membrane is all that stands between a cell and certain death. This continuity is the object of an ongoing struggle between opposing teams of proteins: a proapoptotic squad, featuring brutes with names like "Bad" and "Bax," and their antiapoptotic rivals, led by cell savers Bcl-2 and Bcl-x_L. If these defenders bind and stalemate the others, the cell is safe. But if the pro-death team gets the upper hand, they will punch holes in the mitochondrion, releasing the vengeful cytochrome c. The balance of power between the two teams acts as an internal "apostat" and determines the probability of suicide.

A variety of catastrophes and rumors can engage the regulators of apoptosis and influence a cell's decision to live or die. Damage to the cell's DNA, for example, is a surefire way to convince a cell that suicide is the right choice. Injury and infection are others. During development, however, suicide is almost always someone else's idea. With a heartless word or a withering silence, a cell's neighbors can rend its mitochondria, unleash its caspases, and "volunteer" it to sacrifice its own life for the benefit of the body.

It was BMPs in the developing hand with a caspase. Read the clues for yourself: BMP2, BMP4, BMP7 in plain view at the scene, the third and final element of the limb (the "autopod," developmental biologists call it), just before the collective suicide of thousands of cells. A ring of death surrounding a bead broadcasting a BMP signal. One hundred percent survival when cells are shielded behind a BMP antagonist. Oh, the accused will try to defend their cruelty by arguing that without tough love, hands (and feet) would be flippers, the digits embedded in a web of connective tissue. This works for ducks, they'd point out, but on land an animal that needs to run, climb, or grasp appreciates independent fingers, opposable thumbs, and bendable toes; either the extra cells wadded into the interdigital spaces or fine motor control had to go. They'll be sure to mention how they respected the cries of "Noggin, noggin!"—the BMP antagonist—coming from the cells busy laying down the cartilage template for future bone. And if all else fails, BMPs will undoubtedly try to shift the blame to others. "FGF made us do it!" they'll whine, "It took away our Wnt!" Admittedly, there is a grain of truth to their argument. Modified by a Wnt signal, BMP actually means "build" (cartilage, that is). FGF, whispered by the apical ectodermal ridge, induces a Wnt antagonist; without Wnt, "BMP" means "go kill yourself." If not for FGF's meddling, the BMPs will protest, they would have helped others rather than hurting them. Regardless, we'll be inclined to acquit them—at least those of us who would rather hold a pencil or play the piano than paddle around a pond.

In the hand, suicide is a response to an explicit request. But the most common reason cells kill themselves during development is not because someone tells them to die but because no one tells them to live. Metazoan cells depend on their neighbors for their very lives, their loyalty to the organism enforced by an inborn self-destruct mechanism programmed to go off unless it is continuously repressed by social interaction. Cell suicide "is a social phenomenon. . . . Cells commit suicide unless signaled by other cells," says developmental biologist Martin Raff. These "survival signals," many of them the same growth factors directing cell division and development, act as a safety lock on the suicide machinery. As long as a cell receives a steady supply of the signal from its neighbors, its weapons are disabled. Deprived, it automatically commits apoptosis.

Growth factors can double as survival factors because the modular construction of signaling proteins and the flexibility it confers allow cells to complete sentences beginning with these signals in more than one way. Signals like FGF that bind to receptor tyrosine kinases, for example, can engage the enzyme phosphatidylinosotol-3-kinase as well as the MAP kinases. "PI3 kinase," as it's often called, is an effector, analogous to adenylyl cyclase in the β-adrenergic receptor pathway. Like the cyclase, its activation leads to the production of a second messenger, in this case a multiply phosphorylated lipid nicknamed "PIP3." PIP3 does a little kinase tinkering of its own. This results in the inactivation of proapoptotic proteins and the empowerment of antiapoptotic counterparts, ensuring that cytochrome c stays locked up in the mitochondria and the cell stays alive as a consequence.

Thanks to another contingency plan, no embryonic tissue is in greater need of a suicidal overhaul than the semicomplete nervous system. Evolution could have equipped the brain to meet the challenge of wiring billions of connections, without gaps, redundancies, or mistakes, by generating exactly the right number of neurons, each guided precisely to the correct target by a wiring diagram written in the genes. Instead, it opted for a monumental "success-through-excess" strategy. Even after Notch and Delta have restricted the number of neuronal precursors, the developing nervous system is flooded with a glut of baby neurons, only half of which are actually needed. Unless the extras are culled—quickly—the brain will pop right out of the head like rising bread dough.

Apoptosis harrows the overgrown nervous system, pitting newborn neurons against each other in a battle for survival. Each must locate the source of an essential growth factor—its designated partner—as soon as possible or face certain death by apoptosis. Competition simplifies the complicated task of assembling the brain's intricate circuitry. Neurons that hook up correctly and promptly are rewarded with enough growth factor to disarm the suicide program. Those that dally are likely to find that the early birds have sopped up the entire supply, while any careless or confused enough to wander off course will find none at all. The strategy is exhaustive—no neuron is likely to go without a partner—as well as self-correcting, because neurons that make mistakes are automatically eliminated.

The most famous of these neuronal elixirs is nerve growth factor, life giver, caretaker, and muse to neurons of the sympathetic nerves, that branch of the peripheral nervous system that aids adrenaline in motivating the body during an emergency. Discovered in the 1950s by Rita Levi-Montalcini and colleagues Viktor Hamburger and Stanley Cohen, "NGF," like other growth factors, binds to a receptor tyrosine kinase called TrkA (and several other receptors as well). Building the rest of the NGF sentence, however, poses an additional challenge, unique to the neuron: its shape. TrkA receptors are concentrated in the growing tip of a long, thin fiber, the axon, which reaches from the point of contact between nerve and target organ to the neuronal cell body, located in a cluster, or ganglion, close to the spinal cord—a distance measured in centimeters rather than nanometers. In order to turn off the apoptosis program, news of NGF binding must travel from the receptor all the way up the axon. Originally, scientists believed that the message was hand carried, by sealing both signal and receptor in an envelope of membrane and shipping this vesicle back to the cell body. More recent evidence suggests that lugging NGF halfway across the nervous system is unnecessary. Even the receptor may not have to bother with a long and tiresome journey. Instead, the TrkA receptor that binds NGF may either pass the signal on to another TrkA receptor inside the cell, which in turn passes it to another receptor and so on, creating a "wave" of TrkA activity from one end of the axon to the other, or designate the PI3 kinase to act as the courier.

Growth factor dependence explains why removal of the apical ectodermal ridge truncates a developing limb, says Cliff Tabin. Critical to the regulation of outgrowth and patterning, FGF4 and FGF8 secreted by the ridge are also essential to survival: cells in the progress zone depend on these factors to dispel thoughts of suicide while they're making decisions and laying down templates for bones. If the ridge is cut away, cells lose the life-giving FGFs and kill themselves in despair. That, argues Tabin, not the demise of some sort of clock marking time spent in the progress zone, is what gives you abbreviated limbs. "In the truncation experiments, cells do not change; they die (by apoptosis). How much of the limb you get depends on how much hasn't died."

To us any suicide is a tragedy; the suicide of those so young they have not even had a chance to live is heartbreaking. But during embryonic development, cellular suicide is commonplace, and death is not simply a fact of life, but is essential to life.

A FEW WELL-CHOSEN WORDS

Specialization is an emergent property of larger societies, human or cellular. But as Harvard's Marc Kirschner notes, the evolution of specialized skills went hand in hand with organization. "There would be no point in having lots of differentiated cells if you couldn't organize them in some special way. And complex cell organization only makes sense if you're organizing something different." Why go to the trouble, in other words, of elaborating cells specializing in movement if they were scattered throughout your body rather than grouped into muscles? Conversely, why build limbs if you had no muscles to move them, no bones to support them, no nerves to direct them, or no skin to protect them?

The division of labor and the quest for greater organization it demanded posed the same challenge to all multicellular animals, and all solved the problem the same way: differential gene expression. Using a battery of transcription factors able to bind to DNA and turn genes on and off, eukaryotic organisms could choose the proteins made by their constituent cells and vary their selections from cell to cell to produce a wide variety of combinations from the same genes. What's more, thanks to a hermetically sealed internal environment made possible by the evolution of the epithelium and the discovery that these gene regulatory proteins could be integrated into signaling pathways, organisms could regulate gene expression from outside the cell as well as inside. Beginning with one cell—the fertilized egg—the clever manipulation of transcription factors and signals enabled the metazoan embryo to alter gene expression again and again, dividing the embryo into progressively smaller compartments of like-minded cells and relying on the spatial relationships between compartments to guide the differentiation of discrete populations of cells dedicated to the performance of a single specific task.

These discussions, which guide embryonic cells through their travels and career choices, contain some of the most ancient and most familiar words in the language of life. Bodies grew larger and more ornate, internal organs more sophisticated, but the vocabulary of development remained surprisingly concise; where organisms might have resorted to verbosity, they chose creativity instead. "The embryo inherits a rather compact 'tool kit' and uses many of the same proteins to construct the heart, the kidneys, the teeth, the eyes, and other organs," writes Scott Gilbert, an argument supported by molecular biologists Mark Ptashne and Alex Gann: "Genes expressed during formation of one part of the organism are often expressed during the formation of other parts as well. What distinguishes hands and feet, for example, is not solely (or even largely) the expression of different genes, but rather expression of common genes, but at different times, in different places, and in different combinations." Recycled within organisms, conserved developmental mechanisms were also used over and over again by different organisms, each adapting the tool kit to suit its specific needs.

For example, compared to the palace of the human body, containing somewhere between 200 and 300 different types of cells (about 60 trillion altogether), the body *Hydra* has built is no more than a nomad's tent, a shepherd's yurt. Yet even this lowly creature, which can circumvent sex and simply bud like a clump of daylilies, can mouth a Wnt sentence—and use it to guide development. (Speaking of daylilies, scientists have discovered molecules resembling β-catenin, the endpoint in the Wnt signaling relay, in plants. This pathway, apparently, might well be so old that to be a multicellular eukaryote is to be familiar with it.) We use Wnt to differentiate the top of the hand from the palm, *Hydra* to differentiate the top of a bud from the base. Similarly, a fly looks very different from a frog, inside as well as outside. The fly's nerve cord runs along its belly; the frog's spinal cord is in its back. The fly has no lungs or heart; the frog has neither wings nor antennae. Yet the embryos of both know many of the same words and tell familiar stories during the course of rearranging germ layers or mapping out appendages. In the amphibian, bone morphogenetic proteins and antagonists like Chordin and Noggin compete for control of the ectoderm; in the fly, Decapentaplegic—a homologue of BMP4—runs headlong, as it diffuses southward, into Sog—a Chordin-like BMP antagonist, a collision that claims part of the ectoderm for the future nervous system. Frog legs are built by Hedgehog and Wnt, fly wings by Hedgehog and Wingless. In both frog and fly, struggles between Notch-bearing and Delta-bearing cells separate the neurons from the nobodies.

Shakespeare used a vocabulary of just under 30,000 different words to compose his plays. Using only a few dozen words, metazoan embryos compose bodies that are also great works of art. The vocabulary of development—concise, adaptable, and powerful—stands as a tribute to how much can be accomplished with a handful of conserved patterns and a few well-chosen words.

"Hey, can you really do that?" Jenny asks. I don't usually leave half-finished drafts lying around where family critics can peruse them, but last night I forgot to clear the kitchen table before calling it a night.

"Do what?"

"You know, grow two heads."

"Well, in theory. Actually it's just a laboratory experiment. But would you really *want* two heads?"

"Hell yeah!" she exclaims. "That'd be awesome!"

LIFE IN THE BALANCE

The weather report called for snow, followed by freezing rain, but, hey, that's why you bought an all-wheel-drive vehicle in the first place. Still, you hadn't counted on leaving the office so late, adding poor visibility to your difficulties. Clutching the steering wheel as the car hovers just above the road surface on a scrim of ice, you remind yourself how *Consumer Reports* extolled the safety record of this model. Suddenly, the minivan in front of you glides sideways. Your antilock brakes grind and grasp for a purchase. Only some feral instinct keeps you pumping the brake pedal and steering into the skid, so that you slide instead of spin. Until you're in your own driveway, your heart's in hyperdrive, you're sweating like it's July instead of January, your muscles have their own antilock brakes.

It's easy to be dispassionate a few hours later, after you've had a cup of tea and warmed up, to tell your sister who called to see if you were safe that the trip was no big deal and sound as if you mean it. Nearly capsized by anxiety during the crisis, heart and lungs now clock along at their familiar steady pace, righted by some internal peacekeeper with a faultless sense of balance.

What impressed physiologist Walter Cannon most about the mammalian body was this inherent stability, that even though assailed by "the fell blows of circumstance"—winter storms and summer drought, accidents and predators, bodily injury (or the threat of it), hunger, thirst, overexertion, infection, and toxins—it was a model of constancy. "Confronted by dangerous conditions in the outer world and by equally dangerous possibilities within the body," he asserted in his 1929 book, *The Wisdom of the Body*, mammals "continue to live and carry on their functions with relatively little disturbance." Rather than simply hoping for the best, they have evolved mechanisms that allow them to resist stress by reacting to it. Cannon called this balancing act—"the coordinated physiological processes which maintain most of the steady states in the organism"—*homeostasis.*

Such stability, Cannon observed, is possible only because biological processes are so elastic. Often misconstrued as a sort of clamping mechanism that locks parameters like body temperature and heart rate at some constant value, homeostasis is actually a dynamic process. Pushed off center, the body pushes back, allows one parameter to deviate from the mean in order to return another to the range of acceptable values. But these corrective responses themselves must also be checked or they will push conditions too far in the opposite direction, placing a new set of excessive demands on delicate molecular machinery. An essential feature of homeostasis, therefore, is an intricate network of checks and balances meant to contain responses to stress before they spiral out of control.

Your "fight-or-flight" response to a skid on an icy road illustrates this action-reaction nature of homeostasis. Confronted with danger, the body marshals resources needed to fuel a getaway or mount a defense. The heart works harder and faster, an obedient liver mobilizes glucose, muscles tense in anticipation. Once the crisis has passed, however, there's no need for internal organs to work so hard. Now the body puts on the brakes, setting in motion mechanisms to slow heart rate, reduce blood pressure, and conserve energy.

Quick to shift gears, your body, thanks to these braking mechanisms, is equally quick to reequilibrate.

Cannon recognized that such a coordinated response to disruptive events could not occur in a large multicellular organism without the rapid dissemination of information. That this communication must involve the exchange of chemical signals was also apparent. Adrenaline, for example, was clearly responsible for the changes in circulatory function, muscle tone, sweating and shivering, and mental alertness that constituted the response to emotional excitement; deprived of its call for action by disease or experimental manipulation, the slightest stress sent the body into a tailspin. But in Cannon's time, only a few such signals had actually been identified and isolated. As for "what influences the signal and how the signal sends orders to the organs that make the correction," that, Cannon admitted, "must remain a mystery until further physiological research has disclosed the facts."

Research *has* taken the mystery out of the discussions that maintain homeostasis. In the signaling pathways responsible for such mundane functions as the upkeep of tissues and the disposition of nutrients, scientists have not only discovered how "the unstable stuff of which we are composed . . . learned the trick of maintaining stability" but also how an accumulation of petty misunderstandings can throw the body off balance, how a misstep here, a stumble there, can escalate until there's no turning back from a freefall into disaster.

FLESH AND BLOOD

I don't remember why I went back upstairs, but it's a damn good thing I did. The new claw-foot bathtub, which should have been draining into the outflow pipe, was instead draining all over the painstakingly selected color-coordinated ceramic tile floor, the result of a hairline crack in its base. Manufacturing defect? A sudden stop of the delivery truck? One too many turns of the wrench during installation? Months later, I'm no closer to an answer, a refund, a new plumber, or an operative bathtub.

Just add it to the list.

The walls were painted only a few weeks before the bathtub was installed, and they already need a touch-up. The bathroom window frame—along with the six others in the upstairs bedrooms—has warped so badly that wedging a screen into it has become a Herculean task. My rocking chair is about to collapse. Another slat is loose on the deck railing, the freezer in the basement just went haywire, and the washing machine makes an ominous clunk during the spin cycle. It's enough to make you dread Saturdays and wonder if mortgage approval ought to be contingent on opening a direct deposit account at the Home Depot.

Maintaining the infrastructure is an ongoing concern for the body as well as the home owner. Accidents can and do happen, even to the most careful. Skin cells rub off on clothes, wash away in the shower. The lining of the stomach needs a steady stream of new cells to replace those corroded by acid, seared by spicy food, mutilated by powerful enzymes. A red blood cell's days are numbered—at about 120 to be exact—from the moment it reaches maturity. Injuries must be repaired and casualties replaced, or the body will deteriorate as steadily as an aging house.

During development, when the embryo was just a mosaic of equivalence groups, replacing cells lost through accident or injury was easy—survivors simply divided to bring the number back up to normal. In the adult, home repair is the province of experts. Some jobs are carried out by cells with good memories, able to recall and repeat words they learned during development: the endothelial cells lining the blood vessels, liver hepatocytes, fibroblasts. Others may have retired their cell-cycle machinery, but these journeymen, fed the right combination of growth factors or embryonic signals, can still duplicate their DNA and divide to yield two identical daughters. And for everything else, there are stem cells.

Forget botox. The profligates that biologists call stem cells have their own secret for staying young: run away and hide in a place far from the machinations of transcription factors with an eye on your genes. While most cells were finalizing their career choices, therefore, stem-cells-to-be gave up on growing up and stole off to secluded environments, or niches, deep inside nascent organs, where they could postpone differentiation indefinitely. There, they will be permitted to divide all life long. But in contrast to drudges like fibroblasts, the progeny of stem cells are not always identical twins. On average, about half the daughters of stem cells leave the niche and complete the task their mothers abandoned, differentiating into replacements for cells lost by mature tissues. The others stay behind and become stem cells themselves, so that the number of cells in the niche remains constant. The closest we'll ever come to being immortal, this little bit of the embryo in the adult body is a self-perpetuating, inexhaustible resource, ready to be tapped whenever accident or attrition creates a need for new cells.

The skin is an example of a tissue that could keep a contractor employed seven days a week. Shielding the lower layer, or dermis, is a roof, the epidermis, shingled with stacks of interlocking keratinocytes (named for the cords of keratin protein stacked end to end inside of them) and in constant need of repair. For starters it's always losing shingles. The "cells" nearest the surface are actually ghosts; dead keratinocytes shed in prodigious numbers every day. To replace them, the epidermis relies on a cache of stem cells stored near the epidermal-dermal border. Like machine parts loaded onto a conveyor belt, cells continuously exit this so-called basal layer, rise toward the skin surface, mature into full-fledged keratinocytes, and finally die, their internal organelles crowded out by keratin filaments, their future reduced to an afterlife as house dust.

As if planned obsolescence wasn't trouble enough, the skin, our largest organ, is also one of the most vulnerable, enduring "more direct, frequent, and damaging encounters with the external world than any other tissue in the body." Assailed by wind, weather, ultraviolet light, and the occasional sharp object, the skin must marshal stem cells in the epidermis as well as fibroblasts and endothelial cells in the dermis to repair the damage. For these jobs—complicated, with several kinds of cells to engage and direct—the skin needs to be a more articulate conversationalist. Reaching into the past, it must recall words that will fire up the cell cycle and motivate the dormant; billing and cooing, it must recruit and educate the immature.

Conditions at the site of a freshly inflicted wound are no different from those at any accident scene—hectic, theatrical, and noisy, a pandemonium of cries for help, police and paramedics barking orders, onlookers offering help and comforting the injured. Once the damage is contained, however, the reconstruction team is ready to go to work, led by blood cells—actually fragments of larger cells—known as platelets. "You there!" they shout at fibroblasts in the nearby dermis. "Platelet-derived growth factor!" The fibroblast crew snaps to attention, begins dividing, and dives into the wound bed to hammer down floorboards of collagen, the foundation of the tough scar that will replace the spongy blood clot. "Good job," say the platelets. "You've earned a promotion." Now the fibroblasts can issue orders of their own: the fibroblast growth factors FGF2 and FGF7. Then they're promoted again, to myofibroblasts, lectured by platelet-derived growth factor until they are tough enough to contract and pull the edges of the wound together.

Meanwhile, in the basal layer of the epidermis, epidermal growth factor (EGF) and TGF-β (the patriarch of the family that includes bone morphogenetic proteins) talk to the keratinocytes and their progenitors. EGF gives motivational talks to the stem cells, encouraging them to divide. Assisted by FGF, it also orders the new keratinocytes to slide sideways and raise a roof over the wound. TGF-β gives keratinocytes the green light to call in the plumbers. I can't find someone to reconnect a bathtub, but with a single word—"vascular endothelial growth factor," or "VEGF"—the cells rebuilding the skin can command the installation of an entire network of new capillaries.

Generating the blood destined to flow in these vessels is a more complicated project than building skin, however. Blood contains at least 10 different kinds of cells: erythrocytes, the red blood cells that ferry oxygen; three types of phagocytes, white blood cells eager to devour bacteria and scavenge the corpses of fallen cells; B cells that secrete antibodies and T cells that kill damaged or infected cells and nurture B cells; natural killer cells; platelets and the megakaryocytes that spawn them. All are born in the bones, and all are the handiwork of a single cadre of uniquely versatile stem cells.

Their queen arrived without fanfare, hidden under the bark of an ornamental shrub. Instead of accepting a crown, she marked the beginning of her reign by breaking off her wings. Now she lives to lay eggs and will never again see the light of day.

The walls of their very ordinary suburban house, my neighbors learn, have become home to a great dynasty: blind workers, fierce soldiers with huge snapping jaws, pale nymphs on the threshold of adulthood, and at the dark, cool heart of the nest, the queen of this Empire of Termites, reclining in her brood chamber.

The human skeleton is home to another occult kingdom ruled by a caste of immigrant queens. While the bone matrix slowly turned to stone, its substratum of spongy connective tissue, the marrow, remained red, warm, and alive. Earlier in development, hematopoietic ("blood-making") stem cells, born in so-called blood islands—clumps of mesoderm orbiting the yolk sac—poked around the primordial aorta, visited the nascent liver. Now they are ready to come home. Tired of foreigners and suffused with archetypal memories of a faraway palace waiting for them in the bones, they ride the bloodstream to humerus and femur, pelvis and sternum, crawl out of the bone capillaries, and settle in the marrow. Now, like the termite queen in her timbered hall, these hematopoietic stem cells recline on their cushions, breeding blood cells as inexorably as she lays eggs.

Hematopoietic stem cells never entirely give up their wandering ways. Even in adults, "at any given time, a few—in a mouse, maybe a hundred cells—can be found in the circulatory system," says Amy Wagers, a stem cell researcher at the Joslin Diabetes Center in Boston. But the overwhelming majority, she emphasizes, stay in the bone marrow. If they left, they'd have to grow up. "Hematopoietic stem cells need the niche to stay stem cells," says Wagers. "If a daughter cell loses contact with niche cells, it goes on to differentiate."

And what are the words to this spell that confers eternal youth and an infinite capacity for self-renewal? The bone marrow cells kept the secret for a long time, but scientists have finally pieced together a few of the words—and discovered that the best way to talk to infants, even infantile cells, is in baby talk. Some of the same signals that controlled cellular fate and drove cell division in the embryo can be found in the mature organism cooing at stem cells. Wnt, for example, can keep even hematopoietic stem cells transferred from their precious niche to a culture dish dividing and replenishing. Notch is another synonym for "stay young"; Sonic hedgehog is a third.

In times of crisis, these signals can turn up the volume, sending cell division into hyperdrive. At peak capacity the hematopoietic stem cell population may increase by as much as tenfold. "Mobilization is an evolutionarily conserved process and probably originated as a response to catastrophic blood loss," suggests Wagers. Legions of new stem cells rush out of the bone marrow, migrate to the periphery, and colonize sites in the liver and spleen. From these migrants, "you can expand the entire blood cell tree," Wagers notes, adding that as few as five stem cells are enough to reconstitute the hematopoietic system: erythrocytes, phagocytes, granulocytes, T and B lymphocytes, and platelets.

Bearing life-giving oxygen and fighting off enemies are both honorable professions; still the daughters of blood stem cells relinquish their immature obsession with dividing and their antipathy to commitment slowly and cautiously. You can't help but sympathize—to take that first step, altering their gene expression pattern just enough to become a so-called lineage-restricted stem cell, they must give up both their limitless potential and their immortality. They can still divide, and their identity is still a work in progress. But some have committed to the myeloid path, which will train them to become red blood cells, platelets, or phagocytes and others to the lymphoid path, where they will prepare for careers as B cells or T lymphocytes—both irrevocable decisions—and all will die, most within six to eight weeks of that fateful choice.

Each step on the path to maturity entails another sacrifice. The daughter of a lymphoid progenitor must commit to either the B cell or the T cell lineage; the daughter of a myeloid progenitor may continue down the path marked "red blood cells" or the one marked "macrophage," or one of the granulocyte pathways, but she may not cross paths or change directions. Progeny of these committed lineage-restricted stem cells relinquish the last of their stem cell privileges: the right to divide. Resigned to their fate, they can do no more than refine their gene expression pattern to complete the task of preparing for their chosen jobs.

Such weighty decisions for infant cells, so unreasonable to expect them to muddle through on their own. Fortunately, they have a school right on the palace grounds. Cavities and alleyways serve as classrooms, and instructional materials are provided by the stromal cells in the bone marrow, the endothelial cells of capillaries, or mature blood cells. Cells that graduate earn the right to a job in the blood, with a guarantee of lifetime employment.

Classes here at Marrow High are taught in "Cytokine," a dialect related to Hormone. In fact, you can think of cytokines as minihormones, made and secreted by small clusters of cells rather than endocrine glands. Like adrenaline, estrogen and progesterone, thyroid hormone, and insulin, cytokines are words meant for discussions with colleagues who don't live next door, but while endocrine hormones travel far and wide in the bloodstream, most cytokines have a shorter range of influence; they are a middistance medium.

Like growth factor receptors, cytokine receptors relay signals through tyrosine kinases; like G protein–coupled receptors, this kinase transducer is a separate protein. Known as "Janus kinases," after the two-faced Roman god, or JAKs, these transducer kinases respond to the binding of cytokine by phosphorylating themselves and then the receptors. The receptors then phosphorylate an SH2-interaction domain-containing transcription factor of the STAT (for *s*ignal *t*ransducers and *a*ctivators of *t*ranscription) family. Finally, two STATs pair off to form a single gene regulatory complex that carries the message into the nucleus.

Here's the curriculum for a developing red blood cell. A freshman precursor cell of the myeloid (red blood cell/macrophage) lineage signs up for the introductory course "Commitment 101." Required reading: the cytokines stem cell factor (known colloquially as "Steel"), interleukin-3, and interleukin-9. A passing grade entitles a graduate of this class to call itself an "erythroid precursor cell," automatically registered for next term's "Introduction to the Erythrocyte Lineage," where it will delve deeper into the meaning of interleukin-3 and Steel. After that, cells take a sequence of laboratory courses. In "Responding to Hypoxia I and II," they will hear lectures by one of the few cytokines adventurous enough to brave a trip through the bloodstream: erythropoietin, produced and secreted by the kidney. When they're done, red blood cell aspirants will know how sensors in the kidney monitor blood oxygen, how a crisis that causes oxygen levels to fall—a move to the mountains, blood loss from a massive injury—prompts the kidney to release erythropoietin, and how they should respond to that alarm. To round out their education, students will complete an internship in which they'll practice the preparation and use of hemoglobin and a final exam in which they'll be asked to jettison their nuclei and pack themselves with the hemoglobin they've made. Cells that complete the entire course of study will be awarded a "Master of Oxygen Transport" degree, entitling them to begin their 120-day stint as cellular oxygen tanks.

We chose plastic laminate when we replaced the kitchen floor because it looks like wood but is easier to repair. And when someone does drop a kitchen knife or break a glass, we're prepared—we have six boxes of the stuff stored in the garage. When we get around to repainting that bathroom, we have three cans of paint (wall and trim colors) as well as drop cloths, drip pans, brushes, and rollers. Extra wallboard? We've got that, too, stacked behind the flooring. The old curtain rods? Wouldn't dream of throwing them out. We've even bought replacements for those old windows—and enough to replace the windows downstairs, too. In fact, we've accumulated so many building materials in the course of remodeling it's hard to tell if that structure on the side of the house is a garage or a lumber yard.

Apoptosis is the body's alternative to a yard sale. On a day-to-day basis, it regulates the number of stem cells and their progeny so that the number of cells born never outstrips the number lost. Even in times of crisis, when stem cell reservoirs are churning out replacements to repair some catastrophe, it maintains homeostasis and prevents unnecessary clutter. Finally, suicide offers cells a way to atone for any errors made during the course of cell division—an important safeguard when continuous transit through the cell cycle is a way of life.

Silence and isolation can convince any cell it's better off dead, and stem cells are no exception. For example, hematopoietic stem cells cultured without the benefit of stem cell self-renewal signals such as Wnt quickly lose their will to live. The proapoptotic and antiapoptotic factors contesting the integrity of the mitochondrial membrane may also play a significant role in stem cell apoptosis.

Increasing levels of the antiapoptotic protein Bcl-2 can give hematopoietic stem cells a new lease on life, and these survivors are more vigorous as well, replenishing irradiated bone marrow even more effectively than the stem cells of normal mice.

Development does not end when we are born. No matter how long we live, the cells set aside to repair body tissues will never grow up, never stop dividing. Conversely, part of us is always dying. Apoptosis sweeps away the superfluous, preserving the balance calculated over millions of years of evolution.

Notions of expansion are unacceptable to a society ruled by homeostasis. The community of cells that constitute the adult body is a closed society and "has no provision for any process which would be the equivalent of immigration. Nor has it any provision for unlimited growth, either as a whole or in its parts," writes Walter Cannon.

"Indeed, when some cells reproduce themselves in an uncontrolled manner they form a malignant disease, endangering the welfare of the organism as a whole. Against such pathology," he concludes, "the body has no protection."

BAD, MAD, OR OUT OF CONTROL?

Question: What do a one-eyed lamb and some cancer patients have in common? Answer: Both have abnormalities in Hedgehog signaling to thank for their problems.

Fresh greens, cool mountain air, and the freedom to wander at will—what could be healthier for a mother-to-be? Sheep ranchers know, however, that in the Rocky Mountain meadows, danger can be as close as the next bite. *Veratrum californicum*, commonly known as the corn lily, grows here, and pregnant ewes that nibble on its broad leaves early in their pregnancies give birth to lambs with misshapen faces and a single monstrous eye. "Cyclopia," veterinarians call it.

In the 1950s a Basque shepherd working for a rancher in Idaho pointed out the corn lily plant to agents from the U.S. Department of Agriculture seeking a reason for a sudden surge in the incidence of cyclopia. It made pregnant sheep sick, he told them. In controlled trials, USDA scientists showed that ewes deliberately fed corn lily leaves gave birth to cyclopic lambs. From the plant's tissues they isolated a cholesterol-like compound responsible for its disastrous effects on the developing fetus and called it, appropriately enough, "cyclopamine."

Cyclopamine caught Philip Beachy's attention because he'd learned that mice lacking the signaling protein Sonic hedgehog were also born with cyclopia. Beachy reasoned that cyclopamine might be interfering with Hedgehog sentences; based on its chemical structure, he thought at first it might be disrupting the synthesis of the all-important cholesterol tail. "Then we learned that it interfered with signaling, not processing," he says. "It blocks cells' ability to respond to Hedgehog, and its target is Smoothened. In animals treated with cyclopamine, the Hedgehog pathway is constantly suppressed, even if Sonic hedgehog is present, because the drug disconnects Smoothened from the transcription factor complex."

Gorlin's syndrome, also called basal cell nevus syndrome, causes a constellation of birth defects that are the mirror image of cyclopia. Instead of a single eye, patients with this heritable condition have abnormally wide-set eyes and a broad, flat nose. But the most disturbing feature of Gorlin's syndrome is not how patients look but how easily they develop cancer. A diagnosis of Gorlin's syndrome is nearly synonymous with a diagnosis of basal cell carcinoma, the most common form of skin cancer. Nearly all patients eventually develop skin cancer—unless they succumb first as children to the brain cancer medulloblastoma.

Cyclopamine silences Hedgehog signaling. In Gorlin's syndrome the Hedgehog signaling pathway is never quiet. "These patients turn out to be heterozygous for Patched," says Beachy (that is, one copy of the two copies of the gene is defective; the other, inherited from the other parent, is normal). Every patient eventually gets basal cell carcinoma when a second mutation in the remaining Patched allele results in a total loss of function." Without a functioning Patched receptor, Beachy explains, "Smoothened fires constitutively. The pathway is full-on, even in the absence of Sonic hedgehog, because there's no repressor."

Some cells just seem to be looking for trouble. Basal cell carcinomas originate in the basal layer of epidermis that includes skin stem cells and their immediate progeny, still immature and capable of dividing. Tissues like these, home to noteworthy populations of stem cells, constantly in need of repair, and among those most likely to encounter carcinogens—chemicals, ultraviolet light—are an accident waiting to happen, poised on a knife-edge between normal repair processes and pathological overgrowth. "Stem cells already look a lot like tumor cells," Beachy says. "They already have the capacity for growth. It's just restricted to a particular niche." A little misunderstanding, an incendiary word, a heritable case of logorrhea and cells like these may escape the homeostatic constraints that ordinarily temper cell division, a failure that can quickly endanger the welfare of the entire organism.

Eyes see, brains think, and lungs breathe. Skin protects; bones support. Cancers grow. Reckless proliferation is cancer's trademark, encroachment its principal weapon in a battle with the body. As cancer researchers Gerard Evan and Douglas Green put it in 2002, "Much of the characteristic pathology of cancer arises spontaneously as a consequence of interactions between the expanding mass and its somatic milieu." Cancer's growth spree often begins with a media takeover: subversion of the signaling pathways that regulate proliferation and embryonic development. The architects of this rebellion are perverted versions of cellular signaling proteins, receptors and kinases that have had an amino acid or two deleted, added, or swapped in a genetic accident, so they no longer mean quite the same thing they did before the mishap.

Mutation is the price organisms pay for a long life in a perilous and imperfect world. Spontaneous chemical reactions, reactive forms of oxygen, ultraviolet light from the sun, and environmental toxins assail the DNA. Retroviruses leave their genetic garbage in the genome. Every time a cell divides, whether it's to build a limb, maintain the blood supply, or heal a wound, a gene can be miscopied or duplicated. Pieces of chromosomes can trade places, separating genes from regulatory sequences. Fortunately, innate surveillance and repair mechanisms spot and correct most mistakes before they spawn deviant proteins. Should a mutation slip past unnoticed, however, the corrupted gene will give rise to a corrupted protein.

Genetically reengineered proteins can disrupt conversations related to proliferation—and trigger a cancer—in two ways. Expansions, deletions, or rearrangements in genes that encode growth factors and developmental signals, their receptors, or components of their intracellular signaling pathways replace normal counterparts with hyperactive imposters that encourage cells to thumb their noses at their neighbors and divide as often as they please. The genes *src* and *fps*, encoding the kinases that led to the discovery of interaction domains and adaptor proteins, are examples of such *oncogenes*, tricksters that cells mistake for legitimate growth signals. Ras, the benign little GTPase we met mediating between growth factor receptors and MAP kinases, is another, mutated in one out of every four human cancers. Mutations in the genes for receptor tyrosine kinases—the epidermal growth factor receptor, the PDGF receptor, FGF receptors—crop up routinely in breast, lung, and ovarian cancers, leukemias and lymphomas, and multiple myeloma.

Other mutations stifle the voice of reason. They knock out so-called tumor suppressor genes, encoding proteins responsible for putting the brakes on cell division. Cells with these mutations hear "go" but never "whoa." Gorlin's syndrome illustrates the way a deleted negative can drive cells crazy. Without the Patched receptor to keep Smoothened in check, a Hedgehog sentence always means "yes" even if it no longer begins with Hedgehog.

The pressure to maintain homeostasis is strong, and a single mutation may not be enough on its own to overwhelm the interlocking mechanisms that strive to keep cell numbers constant. But the longer a cell lives, and the more often it divides, the more opportunities it has to accumulate mutations, overriding one control after another until it crosses the line into malignancy. Amy Wagers, referring specifically to hematopoietic stem cells, suggests that longevity is precisely why stem cells may be "hot spots" for tumor initiation: "They are the only cells in the blood that live long enough to accumulate all the hits. Even if you get two or three mutations in progenitor cells, those progenitors are going to die."

The evolution of colorectal cancer is a well-studied example of how such a succession of small genetic errors can add up to one life-threatening malignancy. The fourth most common type of cancer in the United States, it typically originates in an outgrowth of the gut epithelium known as a polyp. Small polyps are small problems, from a medical point of view, precancerous assemblages of slightly irregular cells easily removed during colonoscopy. But as a polyp grows in size, its constituent cells become increasingly abnormal, antisocial, and aggressive, finally penetrating the intestinal wall as a full-fledged, invasive cancer. Analysis of the genetic makeup of colorectal tumors at various stages of this clinical progression, carried out by Bert Vogelstein, Ken Kinzler, and colleagues at the Johns Hopkins University School of Medicine, have demonstrated that the progression from polyp to cancer is accompanied by a sequential accumulation of mutations, beginning, in many cases, with the derangement of a critical signaling mechanism—the pathway headed by the embryonic mitogen and morphogen Wnt. Mutations in the gene encoding the protein APC (for adenomatous polyposis coli, a hereditary cancer syndrome in which patients develop hundreds or even thousands of polyps), one of the bullies in the Degradation Complex that attacks the β-catenin protein to keep it out of the nucleus, occur in more than 70 percent of colorectal cancers; look closely and you may find them lurking in even the smallest and least malevolent of polyps. With APC disabled, the rest of the gang disperses. A steady stream of intact active β-catenin molecules chant "go, go" at genes, as if a steady stream of Wnt messages were bombarding the cell. What cells wouldn't divide manically under such pressure?

But the transformation still isn't complete. Spurred on by an oncogene, a polyp can balloon into a mob of delinquent cells yet still remain no more than a local disruption. To graduate to a cancer, the cells must acquire not only a mutation that gives them leave to divide with abandon but also a mutation that allows them to flaunt one final social convention: the obligation to die.

They'll have those expensive Japanese maples you just planted for dinner and enjoy the heirloom tomatoes for dessert. Go ahead and spread blood meal around your annuals—they'll munch as eagerly as if you'd sprinkled salt on the plants instead of a repellant. No fence, even an electrified one, can stop them. Your car can—but you'll pay an average of $2,200 to repair the damage.

Welcome to Pennsylvania, home to more white-tailed deer per square mile than any other state on the East Coast.

At the beginning of the twentieth century, deer were so scarce that the Pennsylvania game commission released 1,200 between 1906 and 1925 to appease frustrated sportsmen. Today, 1.5 million deer are devouring woodlands, farm crops, and ornamental plantings to the ground; colliding with 40,000 to 50,000 motor vehicles each year; and providing sustenance to the ticks harboring Lyme disease. The Pennsylvania Audubon Society estimates that in some areas the deer population is nearly four times what the land can actually support. Even urban parks are experiencing a population explosion—in Philadelphia's Fairmount Park, for example, nearly 300 deer crowd into a space big enough for thirty.

The state animal is overrunning the state because most of the natural barriers to overpopulation have failed. Well-fed deer have more babies. With farmers and home owners serving up a veritable feast, it's no wonder the deer are breeding more like rabbits, with 90 percent of adult does giving birth to twins or even triplets every spring. But at one time even a bumper crop of fawns would have been cut down to size by the forces of nature. Some would have lost out in the competition for food, while many others would have become food, a feast for hungry predators. Now, with food available year-round, wolves and mountain lions decidedly scarce, and hunting curtailed by anxious residents of housing developments that encroach on the forests, the deer have few enemies to keep their numbers within bounds. Once just part of the ecosystem, they are now its greatest enemy, threatening to destroy their native habitat as well as corn crops and landscaping.

"The cancer cell is a renegade," writes oncologist Robert Weinberg. "Unlike their normal counterparts, cancer cells disregard the needs of the community of cells around them. Cancer cells are only interested in their own proliferative advantage. They are selfish and very unsociable."

Tony Pawson is more forgiving. Cancer cells are mad, not bad, he says; they act crazy "because they hear voices that fool them into growing. They think they are responding to an external voice, and so they're behaving inappropriately." Mutations in growth-regulating mechanisms have cursed them with delusions of grandeur. Oblivious to time-honored social principles like cooperation and self-control, they overrun neighbors they no longer recognize as allies.

But to Gerard Evan of the University of California San Francisco Cancer Center, cancer cells, like the white-tailed deer eating my neighbors out of house and home, are out of control because chance and circumstance have upset the balance of nature. Proliferation is only one side of the equation; after all, Evan points out, "our cells divide by the zillion every day": during embryonic development, as the egg expands into a multicellular newborn, and in adulthood, to replace the cells we lose in the course of a long, active life. With so many opportunities for cell division to run amok, cancers should be as common as colds.

Yet rebel cells fail along the way to cancer far more often than they succeed.* The reason, Evan argues, is that under normal conditions the human body, like an undisturbed ecosystem, is a self-correcting arrangement. In the forest, a lucky break—a mild winter, a parasite die-off—is usually balanced by inevitable misfortune—a dry summer, a new plant disease. In the body, cell numbers are kept relatively constant by a similar network of checks and balances. If more cells are born, more must die. Only an unusual conjunction of circumstances can tip the balance.

"Replication is an amazing piece of evolutionary conjuring. It's the easiest thing in the world to proliferate, but it's quite difficult to proliferate incorrectly," Evan says.

The magic lies in skilled locksmithing. "How do you make it easy for you to get into your apartment but almost impossible for someone else?" he asks. "You install a combination lock. To open the door, you have a key that flicks all of the pins, in the correct order. You can't pick the lock by flipping just one pin. You need the key that flips them all." Similarly, he argues, cells discourage illicit building programs by placing growth under lock and key. The pins of this lock are a network of intermeshing signaling pathways that link cell division and cell survival. Thanks to proteins that cells can use in more than one sentence, signals that order cells to grow also command them to commit suicide. In this ecosystem, apoptosis preys on the exuberant, sweeping through a field of proliferating cells and disposing of the extras like a pack of hungry wolves.

When net growth is permissible—to repair an injury, for example—cells have a key that allows them to "get into the apartment" safely: a unique combination of signals able to activate the growth program while simultaneously blocking the death program. "The two together give you something you don't have singly," says Evan.

---

*Our perception that cancer is an inevitable consequence of mutation is distorted, Evan points out, by the fact that we see only the successes.

Coupling proliferation to apoptosis was a master stroke for preserving population homeostasis, a way to underwrite the upkeep on a relentlessly eroding body while minimizing the risk of cancer. As long as birth and death remain tightly linked, even a mutation in an oncogene cannot ignite a population explosion because cells die almost as quickly as they divide. But if cells also lose their will to die, the stage is set for an ecological disaster. A mutation that promotes growth and a mutation that prevents death—that's the "minimal platform" on which all cancers are built, Evan concludes.

"Cancer cells, by and large, don't do anything normal cells wouldn't do following injury or during development," he says. "What's different is how they regulate—or don't regulate—that behavior. They have mutations that make them live out of control."

For centuries, religious law banned suicides from church graveyards; civil law called for their corpses to be dragged through the streets, hanged, or burned. But even the most orthodox wavered when an individual chose to die in defense of the faith, the crown, or another human being. As one sixteenth century writer observed, "If it were not permitted to expose our lives for the lives of others, the physician could not exercise his art in time of plague, nor could the wife brave the danger of contagion to care for her husband stricken by the plague, nor could a shipwrecked man cede to another the plank that is his salvation." The altruistic deaths of heroes and martyrs argued that choosing to die could be a noble sacrifice as well as a grievous sin, that while a man could kill himself out of wickedness, madness, or despondency, in an act of unbelievable devotion, he could also "lay down his life for his friends."

A cell with a damaged genome has an obligation to lay down its life before it becomes the ultimate biological weapon. Like the sacrificial deaths of saints and heroes, such suicides are acts of altruism, undertaken to safeguard the health and well-being of the body as a whole. But should a troubled cell begin to entertain the idea of sedi-

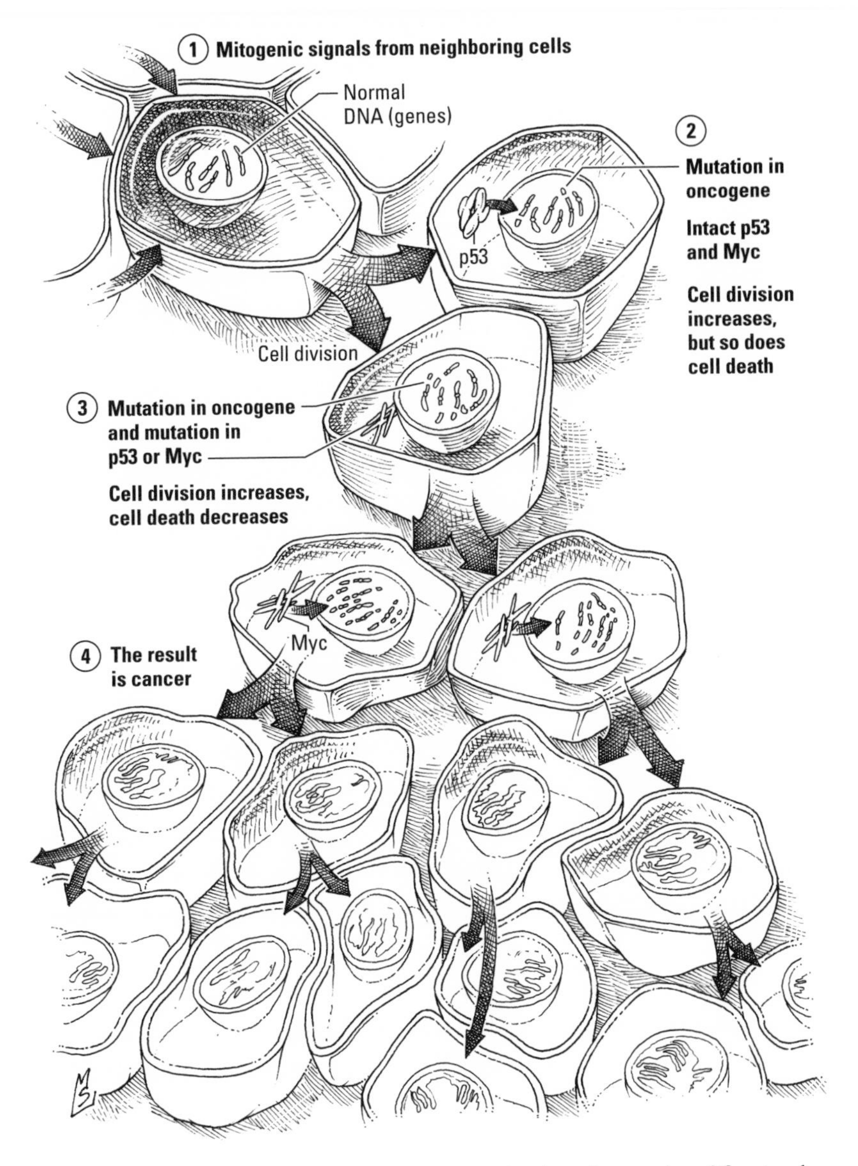

The "cancer platform." Mutations in oncogenes stimulate abnormal proliferation but are not by themselves sufficient to cause cancer because interlocking signaling mechanisms that oversee cell growth and voluntary cell death, or apoptosis, ensure that any signal triggering an increase in growth will trigger an increase in cell suicide as well. A mutation that uncouples these processes or blocks apoptosis, however, eliminates this natural brake on uncontrolled growth and opens the door to a full-fledged malignancy.

tion instead of suicide, a counselor is on call at all times, ready to remind it of its sworn duty to protect the community at large. Named "p53" (for its molecular mass), healthy cells are barely aware of its existence, for most of the p53 protein a cell makes is marked for the garbage disposal almost as soon as it's off the ribosome. When a damaged cell is detained at a cell cycle checkpoint, however, p53 is phosphorylated and stabilized. Now this thanatologist can either shut down the replication machinery or select a go-between (the proapoptotic thug Bax is one of its favorites) to remind the cell of its civic duty to die.

If a mutant cell should slip past the sentinels, the body can still depend on oncogenes. Oncogenes? The corrupted proteins that may have caused that cell to go haywire in the first place? That's right—because many oncogenes are also "apogenes," telling cells to grow one minute and to kill themselves the next. "When you pick up the proliferation program, you pick up the death program as well," Gerard Evan explains.

The most notorious two-faced double-dealer of all is the transcription factor biologists call "Myc," activated when growth factor sentences take a 90 degree turn at the level of Ras, opting for the clause with the second messenger PIP3 as its direct object. A real workaholic, Myc oversees the expression of nearly 10 percent of our genes, from metabolic enzymes to structural proteins, from proteins needed to copy DNA to the proteins operating the cell cycle. In its active form, therefore, Myc can fiddle with switches, pull out plugs, and reprogram machinery all over the cell. For example, at the same time it's cranking up cell division, Myc is also turning in the dividing cells to the police. "Just trying to help," it insists, as it adjusts a cell's hearing—the better to hear signals of stress and injury certain to trigger suicidal ideation. "How about some air in here?" it chirps, as it opens windows in the mitochondrion, releasing deadly cytochrome c. Perhaps it would caution cells not to run with scissors, but these scissors are caspases, and they cut so beautifully.

"Myc has these two contradictory properties—it drives cell proliferation and it drives cell death. And that seems counterintuitive," notes Evan, "until you compare it to putting the brake on when you start your car. Before you shift into drive, you first need to protect against accidents. You take a second look to see if it's safe to proceed before you actually go anywhere." Similarly, he continues, Myc reminds the cell to "stop, look, and listen" before it dives headfirst into proliferating. "Myc equips the cell with the wherewithal to expand and remodel and, at the same time, with all it needs to abort if necessary."

Mutations in Myc are tantamount to brake failure. A cell with a mutated oncogene is already in a big hurry to divide. Without Myc to insist on caution, it shoots out of the driveway and peels away full-speed toward disaster, running over nearby cells unable to get out of its path.

Apogenes and altruism are for conformists. A cancer cell obeys no social conventions, accepts no limits, acknowledges no authority. Bad, mad, and out of control, even a direct command to commit suicide may no longer contain its pathological rebelliousness.

My husband pounced on the box of books at a yard sale, paid five dollars for the lot, and dragged them home. "I had these as a kid. We bought them in the grocery store," he mused as he took them out one by one and read their titles: "*Mathematics. The Insects. The Universe. The Planets.*"

"Hey, give that here," I interrupted, recognizing his find immediately as a nearly complete collection of the *Time-Life Nature and Science Library*. "I remember these. They were one of the things that got me interested in science."

Vishva Dixit grew up in a rural village in Kenya, far from a grocery store, but he too was attracted to science by the *Time-Life Nature and Science Library*. "I can't imagine it now," says Dixit, now a senior researcher with the South San Francisco drug company

Genentech, "but we used to get these *Time-Life* science books, called things like *The Insects* or *The Planets*, 10 miles from the equator, in an African village. I looked at them, and I don't know if it was a curse or an enlightenment, but that just captivated me. I said, 'This is what I want to be.'"

Dixit's parents wanted him to be a doctor—a "profession that guaranteed security"—not a scientist. He completed medical school in Kenya, then chose "an enlightened research-oriented program" at Washington University in St. Louis for his residency. Six years later, Dr. Dixit had his own laboratory, as well as his medical degree, and he'd just discovered a problem he couldn't wait to investigate.

"I ran into an article in *Scientific American* on tumor necrosis factor, this protein that could literally make tumors melt away," he recalls. "And it struck me that here was a substance that you just sprinkled on cells and, lo and behold, they died, while you observed them, in a matter of hours. They were as dead as doorknobs. There was nothing complex. This wasn't a mouse you had to manipulate or a dissection you had to do. The question was 'What happened?'"

Tumor necrosis factor, Dixit had learned, was a subtle chemical weapon. It didn't simply injure or poison cancer cells. It ordered them to kill themselves.

The decision to die is not one that cells take lightly. They would prefer to wait for an explicit command, from p53, say, or from a signal like tumor necrosis factor, one of a family of signaling proteins that apoptosis researchers call "death ligands." When Dixit began his quest to learn how tumor necrosis factor persuaded cells to kill themselves, the scientific community knew that this death ligand, like other signaling molecules, transmitted its sinister message to its victims by way of specific receptors that spanned the cell membrane. They even knew the DNA sequence of the tumor necrosis factor receptor. What no one could figure out, however, was what happened after signal and receptor connected.

"Much to our chagrin, there was nothing in their [the receptors'] sequence that suggested a signaling mechanism. They weren't linked to a G protein. They weren't kinases. How in blazes were they signaling?" says Dixit.

Then scientists studying apoptosis in the development of *C. elegans* discovered the ced-3 protease—and that a mammalian enzyme used in the production of certain cytokines, interleukin-1β-converting enzyme, or "ICE," closely resembled the worm protein. They speculated that an ICE-like protease might be the elusive death effector, a hypothesis that led to the discovery of the first caspase. In addition, Dixit learned that the receptors for death ligands were molecular arms dealers. Like the adaptor proteins of receptor tyrosine kinase pathways, death receptors contain an interaction domain, the "death domain," that enables them to recruit a band of confederates—other adaptor and scaffolding proteins—and link them to form a chain that allows them to access the killer caspases.

"In the end, it was a very simple model," Dixit observes. "What I liked about it was the elegance of its simplicity. It was very chaotic when we began to work on it, a morass of confusion. Yet when the smoke clears, you are left with this work of art."

Actually, recent evidence suggests that tumor necrosis factor signaling is not quite as simple as it first seemed. Just as Myc can promote cell division or call for cell death, this death ligand, it turns out, can also be a life ligand. In fact, the initial reaction of the adaptor that's first to join up with an activated tumor necrosis factor receptor, known as "TRADD," is not to look for a caspase but to call for a kinase that forwards news of ligand binding to the transcription factor NF-κB, an overseer of proliferation genes. The activation of NF-κB then rouses antiapoptotic factors that promote cell survival rather than cell death.

Thoughts of suicide don't surface until this "complex I" breaks up, freeing TRADD and the kinase to drift into the interior of the cell in search of new partners. Now another death domain–containing protein, FADD (for *F*as *a*ctivated *d*eath *d*omain, Fas being another of the death ligands), makes an appealing sidekick. FADD already has connections to a signaling caspase; together with receptor-activated TRADD, the three form a second signaling complex, complex II, able to activate a killer caspase and trigger apoptosis.

In other words, the tumor necrosis factor receptor, more like Myc than one of the proapoptotic thugs that punch holes in mitochondria, is a checkpoint where cells can weigh the decision to live or die. If signaling through complex I and NF-κB wins out, the cell will survive. If the voice of complex II drowns out the proliferation option, death by suicide is the decision. But one little mutation is all it takes to override the voice of reason. Addle the tumor necrosis factor receptor or one of the adaptor proteins and apoptosis drops off the list of options. Now the terms "death ligand" and "death receptor" have new meanings. Without the brake of suicide, cells can plunge into a frenzy of malignant proliferation; worse, the activation of NF-κB allows cancer cells to resist chemotherapy, compromising treatment. Such mutations have been identified in breast and lymphoid cancers and have been shown to promote tumor growth and metastasis.

Cancer cells have even more devious ways of compromising signaling by death ligands. Some lung and colon cancers, for example, lie to receptors to circumvent the death ligand FasL, elaborated by white blood cells patrolling for such miscreants. Under ordinary circumstances, the binding of FasL and the Fas receptor spells certain death. In one study of 35 lung and colon tumors, however, about half had discovered how to make an inactive soluble form of the Fas receptor. This "decoy receptor" bound and sequestered FasL, shielding the tumors from its lethal effects.

Death ligand signaling, "is, of course, an irreversible mode of signaling," Vishva Dixit concludes. "Then again, death is an irreversible event." But by virtue of mutations that allow them to ignore irrevocable calls for suicide, seditious cancer cells can escape the finality of death, or at least postpone it until their irresponsible behavior destroys their habitat.

We all want to live as long as possible; forever sounds good to many. But as cancer researchers like Gerard Evan can tell you, immortality—on the cellular level at least—is incompatible with life. A refusal to die is an invitation to cancer—a triumph for the malignant cell, a loss for the forces of homeostasis, and a disaster for the body.

EATING TO LIVE, LIVING TO EAT

I meant to eat breakfast. But then I cut my finger with the new bread knife, the phone rang, the bagel burned, I spent 20 minutes looking for the checkbook so I could get the credit card bill in the morning's mail, the phone rang, I remembered that I needed to reschedule the orthodontist because Haley needed to stay after school to make up an algebra test, the dogs wanted to go out, and the phone rang again.

I settled for a second cup of coffee.

And I meant to eat lunch—I even grabbed a package of strawberry Pop Tarts on the way out the door, en route to my riding lesson. I looked forward to snacking on them during the drive home. Then my cell phone rang. I should have pulled over, but I was running late, so I sacrificed eating in favor of keeping one hand on the wheel. Then I had to negotiate a road under repair and a couple of bad drivers. By the time I arrived at the school to pick up Haley, I'd forgotten all about food.

"How can you eat these things?" she demanded, holding my Pop Tarts at arm's length. I did have to admit they were a little worse for the wear after an afternoon on the front seat of a warm car.

"Hey, that's my lunch."

"You should eat a real lunch already."

"If they're so awful, why are you eating them?"

"I need to replace lost energy. I had swimming *and* an algebra test today."

We finally split the Pop Tarts. Then we picked up poster board for Haley's science project, dropped off the videos, and replenished the milk supply. We drove back to the barn so that Haley could commune with her pony and picked Jenny up from work. I gave a short lesson in factoring polynomials, solved the crisis of two collies and a library book, did a load of laundry, found a missing pair of eyeglasses, abused a couple of telemarketers, and recited the many reasons why we could *not* buy a third horse. I wanted chicken, but that would have caused another crisis, so I agreed to spaghetti and meatballs and finally sat down to dinner—my first real meal of the day. (My husband was in Boston that night. He had dinner at the Ritz Carlton.)

It isn't easy balancing the needs of two careers, three other people, and four animals with those of one mother. And it isn't easy for my body, either, balancing a 16 hour-a-day demand for energy with my unorthodox eating habits. Yet despite my best efforts to destabilize my internal milieu, my blood glucose levels have remained high enough to keep me from fainting and low enough to keep my doctor happy. More impressively, my body has juggled supply and demand well enough over the long term that I can still wear the Armani jacket I bought before I had children and began to live on Pop Tarts and pasta.

Six months ago, Joe never thought much about what he ate or when he ate it either. He was one of those people who are "never hungry" in the morning. He was usually "too busy" for lunch. "Really, the only meal I ate every day was dinner," he says. "That and maybe a bowl of ice cream with my wife later while we watched television." But that was before a blood sugar reading over 1,000—10 times the acceptable level for a healthy person—landed Joe in the intensive care unit of the local hospital for a week. Now he juggles portion sizes, food selection, medication, and exercise and worries if eating a piece of fruit will give him enough energy to make it through to lunch or add too many calories, if taking a walk will keep his blood sugar within acceptable limits or send it plummeting too far in the opposite direction.

Diabetes sneaks up on people that way. "I should have seen it coming. I had all the symptoms," Joe recalls. "I was thirsty all the time. Of course, if I drank more that meant I went to the bathroom more often. And my eyes. I couldn't understand why my vision was getting worse and worse, even though my eye doctor had just changed my prescription again. For God's sake, my dad even died of diabetes. And I still didn't realize it until it was almost too late."

Steeped in fluid, the cells of the body drift in blissful isolation far from the chaos of the outside world. For Walter Cannon the constancy of this fluid matrix was the key to a "free and independent life," the shield behind which cells could devote themselves to their specialized functions without risking starvation, dehydration, or exposure. "Insofar as the constancy of the fluid matrix is evenly controlled," Cannon noted, "it is not only a means of liberating the organism as a whole from both internal and external limitations, but it is an important measure of economy, greatly minimizing the need for separate governing agencies in the various organs."

Keeping cells well fed is one of the most important duties of the homeostatic mechanisms regulating the composition of the fluid matrix. Mammals, supporting large brains as greedy as a gas-guzzling SUV, must satisfy a relentless demand for fuel. If they were barnacles, flooded with dinner options on every incoming wave, meeting that demand would be as simple as waiting for the sea to wash over them. But they must take their meals when and where they can, and glucose availability rises and falls accordingly. Yet while a hearty dinner staves off one sort of crisis, it creates another, for the glucose that sustains cells can also poison them. High concentrations of glucose trigger fluid retention, ischaemia, and the release of vascular permeability factors, events that are especially destructive to the delicate capillaries supplying the retina, kidney, and peripheral nerves. To be on the safe side, any glucose the body cannot use immediately should be discarded before it causes trouble—but that would be wasteful in light of the effort invested to obtain that fuel. Regulatory mechanisms must find a middle ground. Too little glucose and active cells will not have enough fuel to meet their energy needs. Too much and the body risks irreparable destruction to the vasculature, resulting in blindness, crippling neuropathy, or renal failure.

Communication is the basis of a sound energy policy, emphasizing a balance between starving cells and killing them with kindness, tempering the after-dinner surge in blood glucose and wasting most of a hard-won meal. A central element of this policy is the dialogue between body tissues and the endocrine gland known as the pancreas, mediated by the hormone insulin. Released in response to rising blood glucose levels (for example, immediately following a meal), insulin titrates the concentration of glucose by suspending the mobilization of glucose reserves and commanding muscle, liver, and fat cells to clear the excess and store it as glycogen or fatty acids. "Waste not, want not," it pontificates, proud that an imminent threat has been transformed into a resource the body can draw on to keep cells in times of scarcity.

For the 18 million Americans like Joe who suffer from diabetes (specifically, type 2 diabetes, the form of the disorder responsible for 90 to 95 percent of cases), this dialogue has broken down. The most prevalent metabolic disease in the world, type 2 diabetes increases the risk of heart attack and stroke two- to fourfold, accounts for 24,000 new cases of blindness each year, and is the most frequent cause of kidney failure and nontraumatic limb amputation. When you add up the costs of treatment, rehabilitation, and lost productivity, diabetes costs the American health care system an estimated $100 billion annually. Once a disease associated with middle age (that's why you'll sometimes still hear it referred to as "maturity-onset diabetes"), children—some as young as 4—now account for an increasing proportion of the nearly 1 million new cases diagnosed each year. Worse, the prevalence of the disease is increasing at an alarming rate both here and abroad; worldwide, the number of cases of type 2 diabetes is expected to double over next 25 years.

When communication between the mechanisms responsible for

cell proliferation and the mechanisms promoting apoptosis fails, the result is cancer, a population crisis. When insulin signaling fails, the result is a metabolic crisis. Once under exquisite control, blood glucose levels spiral higher and higher, with devastating consequences for the health and well-being of the body.

INSIDE THE BLACK BOX

Herophilus of Chalcedon might have made it into the history books even if he hadn't discovered the pancreas. Records suggest that this Greek surgeon and anatomist, the beneficiary of a new mind-set free of proscriptions against tampering with the dead, performed the first public dissection of the human body in the fourth century BC. Did those spectators see it, the warty finger of tissue behind the liver, pointing toward the body wall? If so, their testimony has not survived. But we do know that during one of his dissections, Herophilus took note of the structure now known as the pancreas.

Physicians of his time would have been more familiar with the symptoms of diabetes—profound thirst, frequent urination (the name is Greek for "siphon"), rapid weight loss, coma, inevitable death—for descriptions of the disorder date back to 1500 BC. But it would be nearly 2,000 years before anyone connected the pancreas to the condition or before medicine could offer patients more than palliative treatment to ease their suffering. Pancreas means "all flesh" in Greek, and for centuries the name was appropriate. Most considered it little more than a lump of flesh; even the famed Roman physician Galen maintained that it was no more than a cushion supporting the abdominal organs and the large blood vessels coursing through the trunk. Then, in 1642, German physiologist Johann Georg Wirsüng discovered the duct connecting the pancreas and the duodenum, the initial segment of the small intestine. Professor Wirsüng was murdered by an irate student before he could learn more, but investigators who followed up on his observation demonstrated that the duct was a conduit for digestive secretions. The pancreas was an exocrine gland.

And it was also an endocrine gland. In 1869 a student at the Berlin Institute of Pathology named Paul Langerhans took a second look at the pancreas under the microscope. In addition to the large flat cells surrounding the pancreatic ductwork, Langerhans found islands of small round cells that had never been described by earlier investigators. Apposed to the capillaries rather than the ducts, these cells in the "islets of Langerhans," as they came to be called, delivered their secretion to the bloodstream instead of the gut. Like adrenaline, this substance was a hormone; practitioners of the new twentieth-century specialty of endocrinology christened it "insulin" (from the Latin "insula," meaning island).

Remove the adrenal glands and the consequence was Addison's disease. Remove the pancreas of a dog, Oskar Minkowski and Joseph von Mering of the University of Strasburg reported in 1888, and the unfortunate animal developed diabetes. Conversely, patients with diabetes exhibited degeneration of the islet cells, reported Eugene Opie of the Johns Hopkins University School of Medicine. Diabetes, researchers concluded, was an endocrine deficiency disorder; if so, it ought to be possible to treat the condition by replacing the lost hormone, much as hypothyroidism could be treated with extracts of the thyroid gland. In 1921, Frederick Banting and his student assistant Charles Best did, in fact, reverse the symptoms of diabetes in a dog after surgical removal of the pancreas with injections of a crude insulin preparation—proof in principle that the deficiency theory was correct. Within a year the two had successfully treated their first human patient. Insulin quickly became the standard treatment for diabetes.

For many diabetics, insulin was the difference between life and certain death. But as early as the 1930s, endocrinologists discovered that some patients did not respond to insulin. Indeed, when a test was finally developed to measure blood insulin levels in 1960, these patients were found to have insulin levels that were actually abnormally high, rather than alarmingly low. Insulin-deficient patients were usually young, lost weight, and deteriorated rapidly without insulin treatment. Insulin-resistant diabetics were older, had milder symptoms, and were more likely to be overweight than underweight. Their condition often improved if they lost weight and followed a strict diet, even without insulin.

Diabetes, it seemed, was really two diseases, with similar symptoms and similar sky-high levels of blood glucose but different causes. In the classic, or type 1, form of the disorder, the pancreas falls silent. Patients stop producing insulin because the immune system mistakes pancreatic islet cells for dangerous invaders and destroys them. Insulin-resistant, or type 2 diabetes, on the other hand, is an attention disorder, not a deficiency disorder. The pancreas speaks, but body tissues no longer listen. At first, insulin resistance is silent, because the pancreas does such a great cover-up job. In an attempt to restore balance, it secretes more and more insulin—in effect, trying to get the receptor's attention by shouting. Instead, glucose levels keep rising, glucose homeostasis deteriorates, and clinical symptoms of diabetes emerge.

Not surprisingly, the identification of insulin receptors with radioligand studies triggered speculation that type 2 diabetes might be the result of a defect in insulin receptors, which compromised cells' ability to hear and respond to the hormone. However, a closer look at the relationship between insulin binding and insulin activity quickly undermined this notion, says cell biologist Alan Saltiel. Insulin responsive cells, he explains, can lose nearly all of their insulin receptors without affecting their response to insulin. "Cells have far more insulin receptors than they need. You can get full activation by occupying only 10 percent of the receptors on the cell surface. So if you had, say, a 50 percent reduction in the number of receptors, you might expect to see a shift in sensitivity but not in receptor activation. That started a strong emphasis on signal transduction." The problem was that before 1981 scientists knew almost nothing about what happened after insulin bound to its receptor.

In the late 1970s, insulin researcher Pierre de Meyts created a cartoon for the journal *Trends in Biochemical Sciences* to accompany a review article written by Ronald Kahn, who now directs the Joslin Diabetes Center in Boston. "It was a picture of a guy pointing to a blackboard with this picture of a theoretical cell," Kahn recalls. "It showed the insulin receptor and then inside the cell was this black shading and it said 'and then something happens.' At the bottom of the diagram it said 'glucose goes down.' And the guy pointing to the board is saying, 'We've come a long way since our black box concept.'"

Kahn pioneered the effort to pry open insulin's black box. "Up until about 1981, we knew that the receptor was a binding molecule, but we really had no idea about its signaling," he recalls. "Then in 1982 we published a paper that defined the insulin receptor as a receptor tyrosine kinase. And with that we created a mechanism to get the signal inside the cell."

Like other receptor tyrosine kinases, the insulin receptor has a split personality. In fact, it's actually two proteins in one, a pair of α subunits that constitute the binding site, and a pair of β subunits that constitute the kinase. When insulin isn't around to intervene, the α subunits keep their β partners in a headlock, choking off enzyme activity. Insulin binding frees the active site, allowing the β subunits to phosphorylate each other. Active and attractive, the phosphate-decorated receptor has no trouble drawing other binding partners or convincing them they'd look better with a phosphate or two themselves.

Joslin researchers, including Kahn and Morris White, identified the first intracellular substrate of the insulin receptor in 1985, a protein they called simply "IRS (for *i*nsulin *r*eceptor *s*ubstrate)-1," and then a second, IRS-2. A year later they discovered that IRS proteins bound through their phosphotyrosines to SH2-containing molecules—the next components in the signaling relay. "And that, I think, really set the paradigm," Kahn concludes.

The number of insulin substrates, however, kept growing. Then again, Kahn notes, insulin does a lot more than micromanage blood glucose levels. It also regulates the storage, synthesis, and release of fatty acids, supervises the construction of proteins, stimulates cell growth and proliferation, offers hints during development, and extends cellular life span. To carry out all of these jobs, insulin activates multiple signaling pathways, shifting from one to the other by selecting different options from its vocabulary of substrates.

"Now we know that something like 10 proteins are substrates for the insulin receptor," says Kahn. "Almost all of them have multiple phosphotyrosines, which bind many SH2—and some non-SH2—containing molecules. We know that the PI3 kinase pathway is responsible for most of the metabolic actions of insulin and that the PI3 kinase is itself an enzyme composed of a regulatory subunit, which comes in eight forms, and a catalytic subunit, which comes in two. And we know that it [the kinase] activates Akt [another signaling protein] downstream. So if you just look at the metabolic pathways and you consider the receptor, the various receptor substrates, the various isoforms of PI3 kinase, and the molecule Akt, there are literally over a thousand potential combinations of signaling molecules."

The challenge, he continues, is to select the right combination for each task. Just as stem cells must have a key to liberate cell division from apoptosis, cells responding to insulin must unlock a signaling pathway before they can use it. "We used to talk about insulin and its receptor as being like a lock and key. Now I like to use the analogy of a combination lock," Kahn explains. "You literally have to dial 25 to the right, 14 to the left, and 16 to the right. If you dial 35 to the right and 12 to the left, you're doing a lot of the same things, but sometimes that lock just won't open. The whole thing is a much more delicate balance of signaling operations than we would have guessed. That's what gives a hormone like insulin its subtlety."

In the case of glucose homeostasis, the critical combination is the one that relocates glucose from the outside of cells to the inside. "The transport of glucose is the rate-limiting step in glucose disposal, especially in muscle and fat," says Alan Saltiel. "In type 2 diabetes, glucose transport is defective. There's a bottleneck at this step." Insulin still binds and the receptor still adorns itself with phosphate, but the lock has been broken, the key lost, the combination forgotten. The door to glucose transport is closed, and nothing hormone or receptor can say will open it.

If a molecule of glucose was as small and sleek as a molecule of water, glucose could seep into cells. Instead, it has to be carried in on the back of a dedicated transport protein, GLUT4, shuttled in and out of the plasma membrane by its own fleet of membrane-bound vesicles. Until insulin calls, these transporters are stored, in their vesicles, deep inside the cell, tied to a hitching post researchers believe may be the actin fibers of the cytoskeleton.

The sequestration, liberation, and translocation of GLUT4-containing vesicles is a complicated process, according to Saltiel. "The vesicles have to form. Then they're sorted into compartments. They're tethered to a specific location. Eventually they're released, although only some can actually move. These are tracked, probably by the cytoskeleton. They have to cross a barrier of actin at the membrane, find a place to dock, and finally fuse. All the steps have to work together to control the process."

Just freeing the transporter-laden vesicles from their tethers—the step regulated by insulin—is complicated. In contrast to the average signaling assignment, this job requires two combinations, each opening a distinct insulin-signaling pathway. "Both pathways are necessary, neither is sufficient. They are independent but coordinated," Saltiel says. The first pathway follows the sequence outlined by Ron Kahn: IRS protein—PI3 kinase—PIP3—Akt. The second utilizes a different pool of receptors and a different signaling relay and operates out of a unique location: pockets of plasma membrane known as caveolae, a subtype of the membrane lipid aggregates that cell biologists call "lipid rafts." The combination that unlocks this pathway begins with an insulin receptor and the substrate Cbl, trucked in to take advantage of the available phosphotyrosine by a protein called CAP. Once activated, Cbl and CAP hitch a ride on the raft, binding to the lipid-loving protein flotillin. Here, they fiddle with a series of adaptors that flip the last pin in the lock, releasing a go-between that communicates with the GLUT4 vesicles.

Convoy! The garage opens, and a steady flow of GLUT4 vesicles heads to the surface of the cell, picks up a load of glucose, and trucks it off to be stored as glycogen or made into lipid.

Researchers now believe that the attention deficit at the heart of type 2 diabetes is almost certainly the result of a subtle defect downstream of the insulin receptor, a disaffected adaptor or inept kinase that alters the meaning of the sentence just enough to render it unintelligible to the GLUT4 and its vesicles. However, cautions Kahn, it doesn't have to be the same defect in every patient. "What exactly do we mean by the term 'insulin resistance'?" he asks. "Are all the steps resistant? Only some of the steps? Are some steps more resistant than others? Are they more resistant in some tissues? Individuals probably differ. For researchers that creates a problem finding genes. For clinicians it creates challenges for finding a 'universal treatment,' because each patient is different."

WEIGHT WATCHING: AN EXERCISE IN FUTILITY

"You can talk to each other. That's allowed," says Barbara Reynolds. It's the first night of her seminar "Diabetes and You," and Reynolds, a nurse-educator at St. Mary's Hospital in Middletown, Pennsylvania, is trying to help a dozen anxious people recently diagnosed with type 2 diabetes relax. Over the course of the next month, she assures them, she'll be able to teach them to understand and cope with their illness instead of fearing it. "I ended up specializing in teaching about diabetes because of my mother," she tells the group. "She flunked two courses like this—one of them was mine. That was my motive for thinking it was possible to do this better."

Tonight's session will focus largely on background information: what diabetes is, how it's treated, how and when patients should check their blood sugar, when to call the doctor. One week will focus on medication. But most of the time will be devoted to the one issue that vexes people with type 2 diabetes more than any other: food. Most of the people in this room are overweight, and that weight is almost certainly what upset their insulin signaling mechanisms in the first place.

Obesity and type 2 diabetes are best friends. More than 80 percent of people with type 2 diabetes are obese, and many experts argue that the increasing incidence of diabetes can be explained in part by the increase in the American waistline. About 60 percent of the adult population—and 15 percent of children—are currently thought to be overweight. From the standpoint of glucose homeostasis, even a few extra pounds can be detrimental, if they're in the wrong place. Abdominal obesity—the familiar "pot belly"—is especially likely to lead to insulin resistance and eventually diabetes.

For those who already have type 2 diabetes, losing weight is like getting a hearing aid. When patients shed pounds, their response to insulin improves. As a consequence, their plasma glucose levels stabilize, and their chances of developing serious complications like blindness and kidney failure decrease. That's why doctors emphasize the importance of weight loss and exercise, and educators like Barbara Reynolds spend so much time teaching about portion sizes and "carb choices."

Reynolds doesn't like the word "diet." "It makes me think of deprivation," she says. She prefers to call it "meal planning," stressing how thinking of food this way "puts the choice back in your hands. *You* say what you're going to eat." But no matter what they call their calorie counting and apples-for-cookies trade-offs, patients are likely to find shedding excess pounds an uphill battle—and they're not alone. Despite an estimated $30 billion to $50 billion spent each year on diets, drugs, and exercise programs in this country, most overweight individuals trying to lose weight fail; of those who do succeed, about 90 percent will eventually gain it all back.

Dieting is so often an exercise in futility because body weight, like glucose levels, is controlled by homeostatic mechanisms. "Biological factors determine body mass, which is then defended," says Jeffrey Friedman, an obesity researcher at Rockefeller University. "Deviations in weight in either direction elicit a potent counterresponse to resist change."

The defense of body weight and the troubled relationship between weight and insulin action both feature strong opinions from the most unlikely of orators: the fat cell. Long considered no more animated than a bottle of cooking oil, fat cells, scientists have learned, are every bit as talkative as the pancreas or the adrenal glands. Fat cells, you see, are part of the endocrine system, too.

Jeff Friedman started working on appetite because he didn't get his fellowship applications in on time. With a year to kill until he could apply again, the aspiring gastroenterologist took a job with Mary Jane Krieg, a neurobiologist at Rockefeller University who studied the biological basis of drug addiction. Krieg gave Friedman the job of perfecting an assay for an endogenous morphine-like peptide, β-endorphin, and suggested he contact another Rockefeller scientist, Bruce Schneider, an expert on the use of antibodies to track and capture proteins, for some suggestions.

Schneider was interested in the neurobiology of feeding behavior and was using the assays as part of his research on a peptide hormone that was secreted by the gut and acted in the brain. The hormone, called cholecystokinin, seemed to signal when an animal had had enough to eat; the brain, in turn, sent word to stop eating and find something else to do. Friedman was intrigued "by the idea of a biochemical basis for complex behaviors" and wondered if CCK—or the lack of it—might have something to do with obesity. That was how he made the acquaintance of *ob/ob*, the world's fattest mouse.

They were gluttons, these beasts. Weighing three times as much as a normal mouse, *ob/ob* mice, victims of a disabling mutation, had up to five times as much body fat and a prodigious appetite to go with it. Even on a strict diet they stayed fat. Exactly what the protein encoded by the damaged gene did for a living or why its loss turned mice into butterballs, however, was a mystery. Any reasonable observer might assume that the *ob* gene was a recipe for an enzyme critical to fatty acid synthesis or metabolism, except that an *ob/ob* mouse transfused with blood from a normal counterpart slimmed down to a normal body weight, suggesting that the Ob protein was a blood-borne substance—a hormone—rather than a cell-bound enzyme. Perhaps, thought Friedman, it might even be CCK.

The suggestion would have been easy enough to prove or disprove, if only he knew the sequence of the gene. But at the time, a dozen years ago, cloning a gene was not the routine chore it is today. "You have to keep in mind that this was a time when the Human Genome Project was in its infancy," Friedman recalls. Nonetheless, by 1994, he and his colleagues had succeeded in cloning the *ob* gene—and had quickly abandoned the CCK hypothesis as a consequence. The Ob protein—Friedman renamed it "leptin" from leptos, Greek for "thin"—was an altogether different sort of molecule: a cytokine, part of the same family of short-range quasi hormones as erythropoietin and the interleukins that guide the maturation of hematopoietic stem cells. Moreover, it was made neither in a gland nor a stem cell niche but in that bloated repository of unspent calories, the lowly adipocyte, or fat cell.

Fat cells talking? Friedman couldn't have astounded people more if he had reported that pigs, or bluebirds, or hermit crabs could talk, but it shouldn't have come as such a great surprise, he says. "Every cell has connections, with neighbors or with distant cells. The moment you have organization you have connections between cells."

In fact, "it was known for many years that the fat cells secreted a number of peptides, but no nobody paid much attention and no one could really ascribe a function to them," points out Mitchell Lazar, a researcher at the University of Pennsylvania Diabetes Center. What's more, he adds, "The question of how obesity could do this [alter insulin resistance] was difficult to understand when the fat cell was viewed mainly as a place to store energy." The discovery of leptin, however, catalyzed "a major paradigm shift in relation to obesity and the fat cell and in relation to obesity and diabetes. Leptin was proof positive that something made by fat cells, when it's missing and nothing else is different—like in the *ob/ob* mouse—totally changes the physiology of the organism."

But "leptin can't explain everything. There are other molecules out there," Lazar says—other fat cell hormones. One, adiponectin, increases sensitivity to insulin, promotes the uptake and breakdown of fatty acids by muscle, and reduces blood glucose levels. Another, discovered by Lazar, is called resistin. Expressed during the division and differentiation of adipocytes, resistin, as its name implies, triggers insulin resistance; as a consequence, glucose uptake plummets and blood glucose levels rise.

Obesity-induced changes in the production and secretion of hormones like adiponectin and resistin alter the normal give and take between these signals and may explain why body weight and blood glucose levels are so closely correlated. In lean individuals, the pro-insulin action of adiponectin and the anti-insulin action of resistin counterbalance one another, supporting and fine-tuning insulin's message. Studies of overweight laboratory animals and human patients, however, show that obesity is associated with abnormally high levels of resistin and an abnormal insensitivity to adiponectin. The balancing act upset, adipocyte and pancreas, once collaborators, now work at cross purposes, allowing blood glucose levels to soar to pathological levels.

Losing weight is difficult, but it can be done. Just look at Andrea. She lost 175 pounds in a little more than a year—but she had to lose most of her stomach to do it. Three years ago Andrea underwent gastric bypass surgery, a last resort for the obese. After they've had their stomachs stapled or their intestines rerouted, gastric bypass patients can't possibly overeat. The pounds vanish. As Andrea says, "you like what you see in the mirror."

What patients don't like is the bloating, nausea, and vomiting that follow every dietary indiscretion. Eat just a smidgeon more than your shrunken stomach can hold and you better know where the nearest bathroom is and be able to reach it quickly. Ditto if you eat too fast, eat the wrong things (and you won't know what they are until you eat them and pay for it), or dare to have a glass of water with your meal.

Researchers hoped leptin would make such drastic measures obsolete. After all, injections of leptin made the chunky *ob/ob* mouse fashion-model svelte in no time. In the clinic, however, leptin was a disappointment. With the exception of a handful of individuals with a genetic deficit in leptin production (human equivalents of the *ob/*

The defense of body weight. Signaling mechanisms linking the fat cell and brain regions central to appetite control and eating behavior strive to maintain body weight within 15 to 20 percent of a set point that may be biologically predetermined. When body weight—and fat mass—decrease below the set point, fat tissue produces less leptin (*A*). The decrease in leptin stimulates neurons in the brain region known as the hypothalamus, prompting release of the signals NPY and agouti-related peptide (AGRP), which increase appetite and food intake, restoring weight to the set-point level. Conversely, a few extra pounds add up to enough extra leptin to shut down NPY and agouti-related peptide-speaking neurons; instead, leptin binds to receptors on neurons that use the signals POMC and CART. The action of these signals curbs appetite, leading to a reduction in food intake and weight loss (*B*).

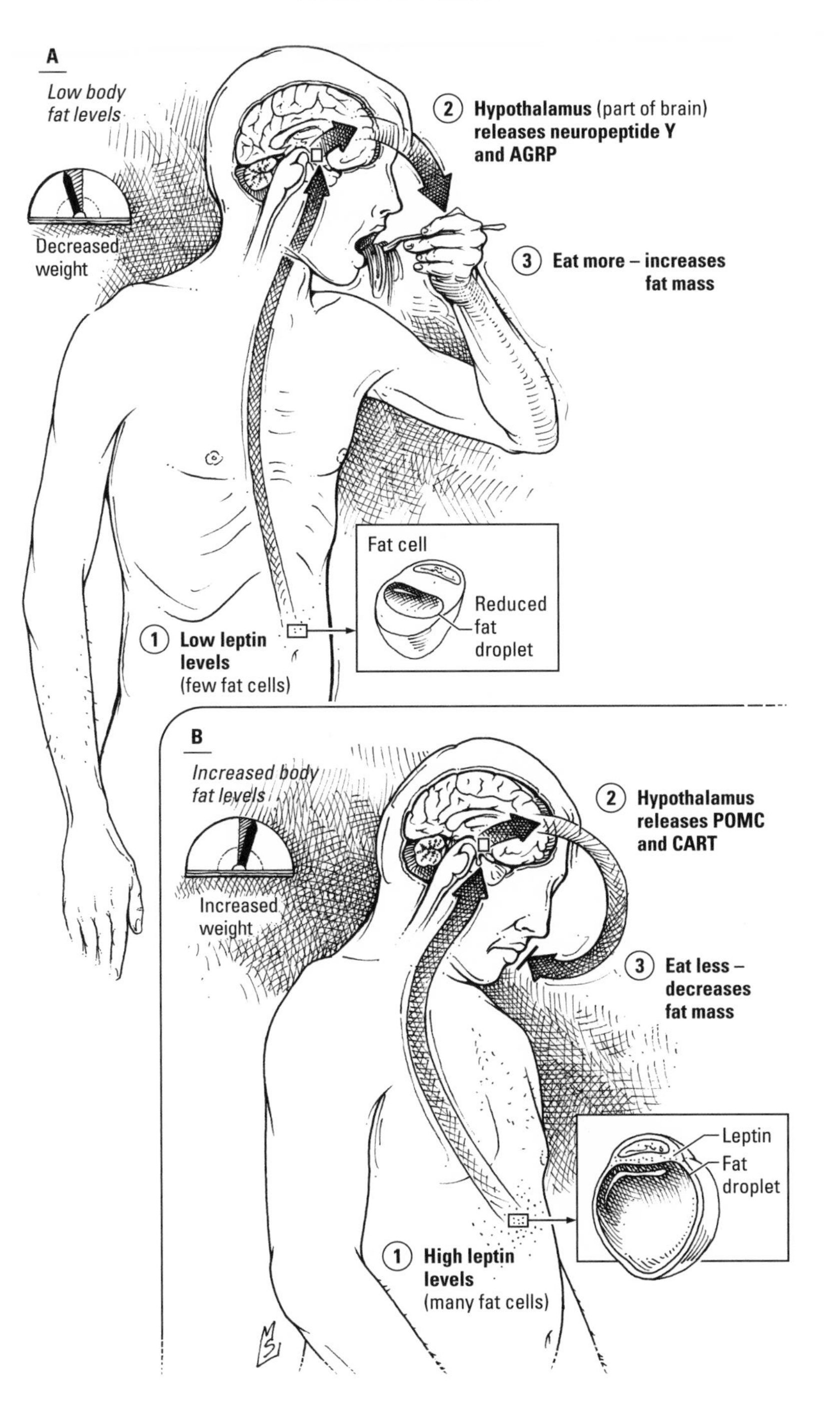
A
Low body fat levels
Decreased weight
2 Hypothalamus (part of brain) releases neuropeptide Y and AGRP
3 Eat more – increases fat mass
Fat cell
Reduced fat droplet
1 Low leptin levels (few fat cells)
B
Increased body fat levels
Increased weight
2 Hypothalamus releases POMC and CART
3 Eat less – decreases fat mass
Leptin
Fat droplet
1 High leptin levels (many fat cells)

ob mouse), obese patients who received the hormone found leptin as ineffective as the dozens of diets they'd tried.

The problem, once again, is an attention deficit. Just as they tune out insulin and adiponectin, the obese are resistant to the effects of leptin as well. Then again, leptin did not evolve to be a natural alternative to fen-phen in the first place. On the contrary, it evolved to keep meat on our bones, and to keep our ancestors from starving when game was scarce, the crops failed, or the cupboard was bare.

Our fat cells worry as much about our weight as we do, and they share their concerns with a most unlikely confidant: the brain, in particular, a collection of cells in the hypothalamus known as the arcuate nucleus. The arcuate nucleus takes a leadership role in the control of feeding behavior, couching its directives, as is the custom in the hypothalamus, in peptides. One team of hypothalamic neurons uses the signals neuropeptide Y (NPY, for short) and agouti-related peptide; they increase appetite and feeding. Their opponents use signals abbreviated POMC and CART; they take the edge off appetite and reduce food intake. The two teams engage in an ongoing tug-of-war over our appetite and food intake, a contest instigated by fat tissue and refereed by leptin.

Arcuate neurons have leptin receptors, allowing them to respond to the concentration of leptin in the bloodstream—Team NPY by giving up, Team POMC/CART by declaring victory. If fat stores decrease, so do leptin levels. Advantage: Team NPY. Result: the individual eats more and gains weight. When fat stores expand, leptin levels go up. Advantage: Team POMC/CART. Result: a loss of appetite and weight loss. In obesity the contest has been rigged. Fat stores increase and leptin levels rise, but no one's listening. Result: brain and body never get the message to eat less, and the pounds stay on.

Feast or famine has been our lot for much of human history, and that past has made the mechanisms regulating body weight really, really touchy. They won't let weight increase or decrease more than

15 to 20 percent beyond the set point without aggressive intervention. And that's why diets fail—as fast as the dieter sheds pounds, body and brain are working as hard as possible to reverse the loss.

"Multicellular organisms store calories as a fat buffer against the time when food is scarce," says Friedman. "Having some fat is adaptive and important." But if banking calories as fat and defending body weight in the face of deprivation are important for everyone, why are some people thin and others so heavy that fat has become a liability instead of an advantage? "The answer, we think, is variation in the system: leptin, leptin receptors, brain peptides," Friedman responds. "When these parameters are less active or less sensitive, the individual weighs more." In other words, people start off with different biologically determined set points, and their bodies then strive to maintain the weight dictated by that set point.

Friedman speculates that the variation in signaling sensitivity and body weight set point may itself reflect the selective advantages of storing more or less fat in different environments. Hunter-gatherer societies, for example, weathered more ups and downs in food availability. "Because they faced the most pressing risk of starvation, it was advantageous for them to store more fat rather than less," he explains. Agricultural societies like those of the Fertile Crescent, on the other hand, "could put away enough grain to last through lean times, so the selective advantage of putting down fat was not present." In fact, he notes, it's a distinct disadvantage to be overweight when the food supply is dependable because obesity increases the risk of gestational diabetes, leading to larger babies and a greater risk of death during childbirth. "There are clearly environments where obesity is adaptive and others where it's disadvantageous. The distribution of weight in the population depends on the premium placed on obesity," Friedman concludes.

Type 2 diabetics can't seem to win. If diet and exercise don't restore glucose levels to an acceptable value, most doctors add an oral medication. But drug therapy has two problems, notes Mitch

Lazar. No drug treatment, no matter how carefully the dose is chosen or how religiously it's taken, even insulin itself, can replicate the exquisite minute-by-minute control over glucose levels achieved by the body. "If you don't treat the disease like the pancreas responds, there are inevitably times when glucose isn't controlled," Lazar explains. Worse, the success you do have comes at a price. Much of the glucose cleared from the blood ends up in fat, aggravating the problem that precipitated diabetes in the first place. "If control is better, patients get fatter. It's enormously frustrating," he says.

The human body has evolved to weather catastrophes, to strive for balance by preparing for contingencies. Unfortunately, a bias toward weight gain that was adaptive when food disappeared at regular intervals is a liability in a society with overstocked side-by-side refrigerators in every kitchen and a fast food restaurant on every corner. Friedman concluded, in a recent review:

> In modern times, obesity and leptin resistance appear to be the residue of genetic variants that were more adaptive in a previous environment. If true, this means that the root of the problem is the interaction of our genes with our environment. The lean carry genes that protect them from the consequences of obesity, whereas the obese carry genes that are atavisms of a time of nutritional privation in which they no longer live.

While developed countries struggle with an epidemic of food-induced diabetes, hunger remains one of humankind's most intractable problems. If you're one of the many stuck in an interminable battle with your adipocytes, you can certainly blame the usual suspects: larger portions, television, carbohydrates, cars. But don't forget to blame drought and disease and war, spoilage and crop failure—and think of the starving. When hunger and malnutrition are only a distant memory for everyone, obesity and its consequences will be as well.

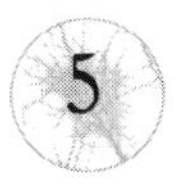

"THE SCENARIO-BUFFERED BUILDING"

All buildings are predictions.
All predictions are wrong.
—Stewart Brand, *How Buildings Learn*

We moved a closet, tore out a vanity, and added a bathtub because the previous owners of our house had installed a dressing room instead of a master bathroom. Across the street, a couple with three growing children bumped out the back wall to extend their kitchen and family room. A more ambitious neighbor raised the roof, added dormers, expanded the garage to include space for three cars, and installed a fence to contain their new puppy. The people on one side just added a deck; those on the other replaced a 12-year-old fence.

"Buildings keep being pushed around by three irresistible forces—technology, money, and fashion," writes author, editor, and designer Stuart Brand in *How Buildings Learn.* Houses and office buildings, libraries and schools, grocery stores and gas stations, hotels, restaurants, museums, and concert halls change and grow continuously, because those who live in them, work in them, and visit them change too. Families shrink and grow. A business that once relied on paper goes broadband. Damage from storms, floods, and fires must be repaired. Trends wax and wane: today, a kitchen island is critical; tomorrow, it's extra closet space.

Unfortunately, most architects, says Brand, prefer to think of their creations as everlasting. The buildings they design resist change. "They're designed not to adapt; also budgeted and financed not to, constructed not to, administered not to, maintained not to, regulated and taxed not to, even remodeled not to." Unconventional floor plans, poor uses of space, rigid materials, and inaccessible plumbing or wiring conspire to make even essential updates difficult and costly. In the most defiant cases, extinction may be easier than renovation, and those responsible for deciding the fate of such an architectural dinosaur may well decide to raze and start over again.

Given that change is inevitable, Brand argues that architects should confront the issue head on instead of avoiding it. Buildings remain versatile, he says, when they're constructed according to a strategy rather than a plan. Strategic or "scenario-based" construction asks those who will use the building to draw up a list of scenarios—ways their circumstances might change in the future—and to examine what the building will have to do to accommodate those developments. Are residents expecting to add to their family? What if a business starts thinking about consolidating divisions, doesn't land a particular job, can anticipate new government regulations? When the possibility of change is actually factored into the design, a structure remains relevant because it can grow and adapt without breaking the bank or running into an insurmountable roadblock. "A good strategy ensures that, no matter what happens, you always have maneuvering room," Brand concludes.

Nothing is certain except change. Whether it is a building or a living organism that will be called on to cope, the longest-lived structures are certain to be those best prepared to accommodate bumps in the road. No one would ever accuse the developer who built my neighborhood of being so farsighted—otherwise he would not have placed a ventilation shaft in an interior closet wall. Evolution, however, is a master of scenario-based building. Organisms must be prepared to adapt or they risk extinction. Since it cannot imagine or foresee the future, however, evolution has instead encouraged mechanisms that confer flexibility and championed processes that allow for experimentation while minimizing the number of fatal mistakes.

The mutability of DNA, the modular nature of proteins, and the co-option of old molecules for new purposes represent some of life's most successful efforts to accommodate the need to adapt. The power and utility of these mechanisms are nowhere more evident than in the evolution of the design elements of the pattern language that made the advantageous transition to a multicellular lifestyle possible. What's more, the signaling mechanisms constructed from these patterns were themselves powerful agents for change. Through the use of structural motifs like interaction domains, organisms could revamp the regulation of conserved biochemical functions, link signaling pathways, and control the timing and location of developmental events.

Yet any mechanism that involves alterations to the basic structure of proteins—even a mechanism as light on its feet as adding and subtracting protein domains—requires the luxury of time, time that generations may have, but the individual does not. In addition, the need to adapt has a personal as well as a global dimension. Each individual has his or her own crosses to bear, problems that are irrelevant to anyone else and that are likely to change from day to day. A truly comprehensive scenario-based strategy for life in a capricious environment therefore had to include ways of adapting in the short term as well as over the long term, a way to accommodate change on a minute-by-minute, day-by-day level within the lifetime of each individual, a solution flexible enough that it could be customized to account for personal experience and unique circumstances.

A seemingly insurmountable problem at first glance, but look—that humble bumpkin, *E. coli,* has its hand up; even it knows the answer. Big-picture evolution has provided the bacterium with enzymes to digest two different sugars. But if the immediate environment is offering an unfamiliar substance and someone introduces a few drops of lactose-rich fluid on the other side of its flask, what's the best course of action? Sit in place and wait for a fortuitous mutation to provide a new digestive enzyme able to deal with the strange food? Of course not—the bacterium stops spinning in place and hitches itself over to the sugar it can already digest. Sometime in the future it may change its genes, but today it changes its behavior instead.

Infinitely flexible and quick as the turn of a flagellum, behavior—a response to a perceived stimulus—was evolution's ultimate coping mechanism, a solution to today's problems and a way to change again tomorrow if necessary. As is true for so many other biological processes—the development of the embryo, the control of proliferation, the maintenance of metabolic homeostasis, the regulation of critical functions like heart rate and breathing—this integration of perception and response (and eventually emotion and thought as well) could never proceed without communication. In the bacterium, discussion is as simple as a chain of proteins, beginning with a transmembrane receptor and ending with the rotor protein cranking the flagellum. However, in metazoan organisms the adoption of a multicellular lifestyle and a gradual increase in size again demanded communication between cells, over longer and longer distances. Hormones were one solution to this new variation on the problem of long-distance message transmission. But hormonal conversations are promiscuous—it doesn't matter which islet cell responds to insulin as long as one does—and conversations coordinating stimulus and response had to be precise. After all, if an animal wants to move its right hind leg, the message must get to the right hind leg muscles, not the muscles of the left hind leg, the right front leg, the jaw, the ribcage, or the tail. Challenged to accommodate the twin demands of scope and fidelity, metazoans came up with a new type of cell, the neuron, with a unique shape well suited to the receipt and transmission of messages over vast distances: at one end, an arbor of short processes, or dendrites, perfect for collecting and integrating information from multiple sources; at the other, a single long process, the axon, that reached out for as long as three yards to contact a faraway partner. To propagate messages from one end of the process to the other, the neuron evolved specialized channel proteins that allow it to manipulate the concentration of charged particles, or ions, and conduct electrical impulses. Much as current flows through a wire, these impulses, known as action potentials, travel down the axon from its origin near the cell body to its terminal.

Axons and ion channels solved the problems posed by distance; a new appliance, developed especially for conversations between neurons, met the need for precision and solved the problem of translating the action potential into a format suitable for cell-cell communication. This structure, the synapse, locked specific neurons into a long-term monogamous relationship. In addition, through the clever use of membranes, actin fibers, and glue, it assembled tools to facilitate the rapid release of chemical signals and sealed off the point of contact, so that neurons could carry on private discussions without worrying that everyone in the neighborhood might eavesdrop on the proceedings. Thanks to these adaptations, cells of the nervous system could use the same type of chemical signaling mechanisms employed elsewhere in the body yet ensure the accuracy essential to the effective, minute-by-minute control of behavior.

In a body of talkative cells, neurons are the most renowned conversationalists. Other tissues discuss the everyday, the practical, or the essential: What should we do about dinner? Am I part of the hand or the head? Could someone please help me patch this cut? The cells that make up our nervous systems describe sunsets, craft poetry, solve equations, remember birthdays, dream. Yet despite their eloquence, they remain true to the fundamental principles of cellular communication. Neurons may sport a college-level vocabulary of perhaps as many as 100 different signaling molecules (neuroscientists call them "neurotransmitters"), but their conversations utilize the same principles and the same signal-receptor-relay syntax pioneered by the simplest organisms, long before the existence of language, thought, or emotion.

"FAR-SPEAKING" CLOSE UP

The year 1876 was a banner year for communication. In Boston, Alexander Graham Bell filed a patent for an invention capable of "transmitting vocal utterance telegraphically," the device that would come to be known as the telephone. And in the Spanish city of Zaragoza, an aspiring doctor named Santiago Ramón y Cajal scraped together the remnants of his army payout and made a down payment on a microscope, an investment that would turn out to be priceless to our understanding of neuronal communication.

Cajal had only a rudimentary knowledge of microscopic anatomy or, indeed, of the operation of his new instrument itself; what little he did know came from a brief tenure in the laboratory of his histology professor. He had never actually prepared a specimen for microscopic examination. He could not read German, so he could not consult the leading textbooks. No one at the University of Zaragoza, where he was employed as an anatomy instructor, knew or cared enough about microscopy to teach him. Too inconsequential to merit laboratory space at the university, he was forced to set up a makeshift lab in an attic. Diffident, socially inept, and more sincere than scholarly, his prospects for a career in academia were modest at best.

Yet what Cajal lacked in technical expertise or sophistication, he made up for in dedication and a passionate commitment to self-improvement. As a boy he had wanted to be an artist, an idea that horrified his father, who apprenticed him to a cobbler rather than permit him to take up such a scurrilous profession. Now, as a scientist, the keen eye, inexhaustible patience, and a matchless ability to re-create on paper what he saw in the microscope were the very traits that made him a sterling anatomist. Gleaning what he could from journals and a few books written in French, he taught himself microscopic technique. To further his study of the literature, he learned German. Applying his newfound skills, he published monographs on the ultrastructure of the epithelium and the muscle fibers of insects and wrote and illustrated a histology textbook. And of all the tissues he observed, described, sketched, or read about in the accounts of other anatomists, none interested him more than the brain.

"In my systematic explorations through the realm of miscroscopic anatomy, there came the turn of the nervous system, that masterpiece of life," he recalled in his autobiography. Here was a tissue truly worthy of his prodigious talents. "To know the brain," he wrote, "is equivalent to ascertaining the material course of thought and will." Yet the difficulties posed by the intricate morphology of nerve cells and the density of their fibers had frustrated the efforts of others to understand the structure and function of the nervous system. That the propagation of nerve impulses involved some sort of electrical activity was clear. But no one could explain how the anatomy, physiology, or arrangement of nerve cells could account for such activity, much less how the brain solved problems or encoded memories. "Nobody could answer this simple question: How is a nervous impulse transmitted from a sensory fiber to a motor one?" Cajal noted.

At the conclusion of his book *Minds Behind the Brain*, in which he reviews the lives and accomplishments of 18 of the major figures in the history of neuroscience, historian Stanley Finger asks what it was that inspired these individuals to greatness. All, he concludes, had a love of learning, an "addiction" to discovery, a healthy skepticism toward established dogma, an "unwarranted optimism." But Finger also emphasizes the importance of insight, timing, and the right frame of mind. "The truly great scientists are observant opportunists whose minds are more open than most to anomalies and new ideas," men and women, he notes, citing physiologist Sir Charles Sherrington, who have "an intuitive flair for asking the right question" at the right time. Perhaps this explains why neuroscientists still speak of Cajal with reverence. Committed, curious, and foresighted, he was prescient enough to recognize that the time had come to ask the difficult questions about the brain, because it might finally be possible, through the studious application of microscopic techniques, to find answers. Bold enough to ask, he was then astute enough to see. "He treated the microscopic scene as if it were alive," Sherrington recalled. Over 100 years ago in a country that was then the most backward of scientific backwaters, using no more than a microscope and his own eyes, Santiago Ramón y Cajal discovered or anticipated some of the most fundamental principles of neuroscience, observations that began to explain how the properties of neurons could lead to the behavior of organisms.

When Cajal began his research in the final years of the nineteenth century, neuroscientists were the only members of the scientific community who seemed unaware that everyone else had embraced the "cell theory"—the idea that cells were the building blocks of living tissues—advanced decades earlier by the German biologists Jacob Schleiden and Theodor Schwann. But who could blame them for their intransigence? In contrast to the simple geometric forms of other cells, the neuron, with its mane of dendrites and its taproot of an axon, resembled something that could as easily have come from the garden as a body tissue. Histologists could not even agree where one neuron ended and the other began. Based on the limited amount of detail revealed by the few available dyes, many argued that the processes of each neuron were fused to those of its neighbors, forming an unbroken network, or reticulum, rather than a conventional tissue composed of discrete cells.

In 1887, Cajal was summoned to Madrid to administer the examinations for doctoral candidates in descriptive anatomy. During his stay, a young neurologist, Don Luis Simarro, showed Cajal some slides of the brain stained with a new technique he had learned in Paris. Building on the well-known reaction of silver salts to light that had inspired the development of photography, an Italian physician, Camillo Golgi, had treated sections of brain tissue with a solution of silver nitrate, hoping to capture more detailed images of neurons. He was successful beyond anyone's imagination. For reasons that are still obscure, the silver stain impregnated a tiny proportion of neurons (somewhere between 1 and 3 percent); in these neurons, however, the stain permeated the entire arbor of processes, revealing the intricate structure of the whole neuron as clearly as if it had been drawn by an artist in pen and ink. Using this *reazione near*, or "black reaction," Cajal learned, Golgi had been able to describe in vivid detail the architecture of neurons in several brain regions, including the cerebellum and the olfactory bulb.

A few iconoclasts had already begun to question the reticular theory, to criticize the possibility that of all the body tissues only the brain and spinal cord were not made up of discrete cells. Surely Golgi's stain would shed light on the matter, everyone thought. But Golgi was the most ardent reticularist of all. Anyone looking at his slides could see how the processes of these cells crossed, overlapped, intertwined—proof, he insisted, that they must be linked seamlessly in a brainwide web.

The silver stain was as capricious as it was beautiful, Simarro cautioned Cajal. Even Golgi himself acknowledged that it was time-consuming and inconsistent, that one experiment could produce the stunning images Cajal had admired while the next failed utterly. But Cajal recognized that in the "black reaction" he had found the "tool of revelation" that would allow him to visualize the brain in the detail necessary to begin unraveling its secrets. He had ideas for improving the technique. And he had an idea why it had still not resolved the question of whether neurons were individual cells or a reticulum.

The problem, as Cajal saw it, was that in the adult brain it was impossible to see the forest for the trees:

> Two methods come to mind for investigating adequately the true form of the elements in this inextricable thicket. The most natural and simple apparently, but really the most difficult, consists of exploring the full-grown forest intrepidly, clearing the ground of shrubs and parasitic plants, and eventually isolating each species of tree, as well as from its parasites as from its relatives . . . the second path open to reason is what, in embryological terms, is designated the ontogenetic or embryological method. Since the full-grown forest turns out to be impenetrable and undefinable, why not revert to the study of the young wood, in the nursery stage, as we might say?

Using a modified, more dependable version of Golgi's silver stain, Cajal probed the "young wood" in the brains of the immature and found that if he applied the technique before the myelin sheath formed, "the nerve cells, which are still relatively small, stand out complete in each section; the terminal ramifications of the axis cylinder are depicted with the utmost clearness and perfectly free." In the cerebellum he not only observed the plump bodies of Purkinje cells described by earlier investigators but also described axons that split and wove baskets around them, as well as axons that climbed the cell body and insinuated themselves throughout the dendritic arbor like "ivy or lianas to the trunks of trees." He observed long processes of sensory neurons coursing into the brain from the retina and olfactory bulb. He saw the elegant pyramidal cells of the cerebral cortex, the chunky interneurons of the spinal cord. Nowhere did he see nerve cells fused together in a reticulum.

From his observations, Cajal drew two historic conclusions. First, the brain, like other tissues, was made up of individual cells, which stood shoulder to shoulder with each other but did not fuse. Second, the arrangement of neurons and the orientation of their processes suggested that the neuron was a one-way street. Sensory nerves, for example, had their dendrites in the periphery, while the tips of their axons were located in the brain—a layout that made perfect sense if the dendritic arbor took in information. The dendrites of motor neurons, on the other hand, were located in the brain or the spinal cord, and their axons terminated on the muscles, out in the body, implying that the axon forwarded information. A neuron, in other words, listened with its dendrites and, if it chose to send its own message in turn, spoke via an inpulse (the action potential) that traveled away from the cell body down the axon.

Sir Charles Sherrington was greatly impressed with the "neuron doctrine" and Cajal's "principle of dynamic polarization"—so impressed, in fact, that he wrangled a speaking engagement at the Royal Society and insisted his Spanish colleague stay at his house afterward.* What's more, Sherrington recognized that if neurons were not joined together and if "the nerve-circuits are valved," there must be something special about the point of contact, some structural feature or mechanism that enabled a signal to cross the gap between axon to dendrite. "In view of the probable importance physiologically of this mode of nexus between neurone and neurone," Sherrington argued, "it is convenient to have a term for it." He called the connection the "synapse," from the Greek meaning "to clasp."

Even the sharp eyes of a master like Cajal could not provide any further detail: the synapse was a gap so narrow it was beyond the resolution of any light microscope. An eyewitness account of the architecture of the synapse—indisputable proof for any diehard who still harbored doubts about the neuron doctrine—had to wait until the 1950s, when neurobiology saw the introduction of a tool even more powerful than silver: the electron microscope.

Like all architectural innovations, the neural synapse evolved to solve problems, in this case the problems associated with distance, preci-

---

*Cajal was an odd houseguest. Not content to stroll in the garden or peruse the collection in the family library, he set up a temporary laboratory in a spare room so that he could continue his experiments.

sion, one-way transmission, and diffusible chemical messengers. Electron microscopic studies of the ultrastructure of the synapse, along with the identification of proteins specific to the presynaptic (the axon) and postsynaptic (the receptive dendrite) membranes, have revealed that this specialized junction incorporates design elements and construction materials developed to address each of these problems. Some, such as receptors that double as ion channels, are distinctly neuronal. Others—specialized adhesive proteins tailor-made to reinforce connections between cells, for example—are adaptations of tools and techniques that metazoans have been working on since the invention of the epithelium, when they first began joining cell to cell in an organized fashion.

By definition the neural synapse is asymmetric. On the presynaptic side the axon terminal is specialized to meet the challenges of directed secretion. Of course, the process of secretion itself is not peculiar to neurons (hormones and digestive enzymes are other examples of substances forcibly extruded by the cells that make them); in all secretory cells the product to be exported is packaged in membrane-bound vesicles, which dock at, then fuse with, the plasma membrane, disgorging their contents into the extracellular space in the process. But the neuron faces a problem unknown to the cells of the gut or the adrenal gland. Proteins, including membrane proteins needed to construct secretory vesicles, are made in the cell body, far from the axon terminal, which may be up to three yards away. An intra-axonal "railway" of microtubules is available to transport these proteins down the axon, but if the presynaptic terminal had to wait for the train every time it needed to get a message across the synapse, moving an arm or a leg might take weeks. To get conversation up to realistic speed, neurons have installed a revolving secretory apparatus at the synapse. Instead of waiting for new vesicles to travel the length of the axon, they recycle the ones they already have, refashioning them from the plasma membrane and refilling them locally. Fully loaded vesicles are then kept at the ready, concentrated at release

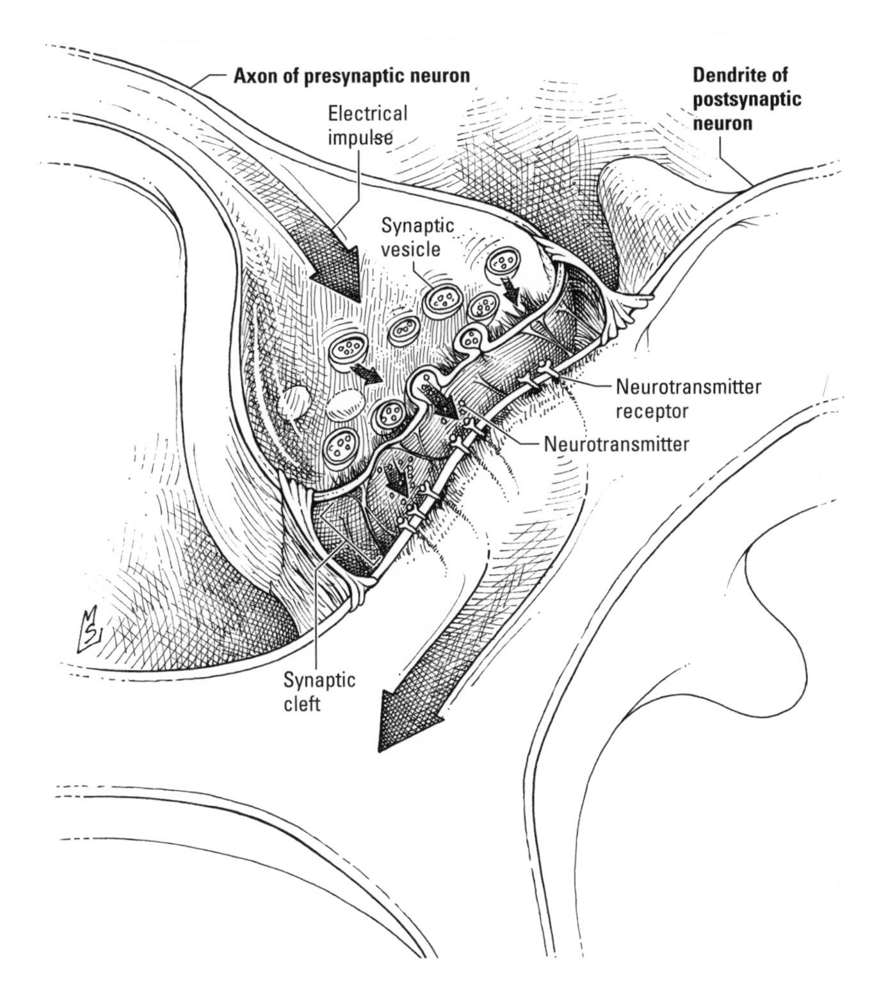

The neural synapse.

sites within the so-called active zone, which looks out directly on the receptors waiting across the cleft. Some vesicles have already redocked at the plasma membrane before the action potential ever arrives; a few have even begun to fuse, poised to expel their neurotransmitter at a moment's notice. Behind these first responders, backup vesicles wait in a grid anchored to the presynaptic membrane, like racehorses pawing in the starting gate. And behind them still more vesicles are held in reserve, tethered to actin fibers in the cytoskeleton and prepared to drop into the grid should a volley of impulses deplete the pool concentrated at the active zone.

The buzzers that spring the loaded vesicles are voltage-sensitive proteins built to channel calcium ions. Planted throughout the active zone, the doors of these calcium channels are kept tightly closed while the neuron is quiescent. The arrival of an action potential flings the doors open, flooding the terminal with calcium. "Go!" it shouts. In response, the vesicles discharge their payload, and the liberated neurotransmitter seeps across the synaptic cleft into the waiting hands of the receptors on the postsynaptic neuron.

Even if you're not a histologist, you'd have no trouble spotting the postsynaptic membrane in an electron micrograph—it's the one underlined in black. Actually a tightly woven matrix of protein, this postsynaptic density, as it's called, locks neurotransmitter receptors in place directly opposite the secretory apparatus of the presynaptic neuron. Some it even arranges like flowers in a centerpiece. For example, receptors for the amino acid neurotransmitter glutamate, which come in two varieties, are organized with those subtitled "NMDA" (because of their affinity for the compound *N*-methyl-D-aspartate) in the center, surrounded by the type known as "AMPA" (for alpha-amino-3-hydroxy-5-methyl-isoxazole-4-proprionic acid, one instance where an acronym is absolutely, totally justified). In addition to receptors, the postsynaptic density contains proteins that participate in signaling relays (kinases, for example), proteins that regulate the internalization and recycling of receptors, and proteins that connect the array to the internal scaffolding of the dendrite.

Holding the entire construct together are adhesion molecules, membrane proteins that reach across the gap in both directions to clasp axon to dendrite like the hooks and loops of Velcro. Within the confines of the synapse, those called neurexins, located on the presynaptic terminal, bind neuroligins, their "receptors," on the postsynaptic terminal. Others, known as cadherins, flank the synapse; their intracellular domains bind to catenins, which anchor them to the cytoskeleton. Adhesion molecules do more than maintain connections and align neurons, however. Available in a wide range of isoforms (variants of the same protein made by splicing mRNA fragments in different ways), the cadherins, in particular, bind one another with a specificity as selective as any signal-receptor pair. During development, neurons rely on these interactions to assist them in identifying an appropriate partner. In the adult brain, adhesive interactions between the "parasynaptic" cadherins form a seal around the synapse that ensures privacy and precision.

The word "telephone" means "far speaking," an ideal term for a device that could allow a person to speak directly with another across town, across the country, or on the other side of the world. Neurons, that felicitous invention of talkative metazoans, are also "far-speaking" devices. With their axons traversing the brain and coursing through the spinal cord, neurons allow one part of the nervous system to talk directly to another, no matter how far apart they are. And where axon and dendrite meet, the architectural features of the synapse direct the release of neurotransmitter, focus the attention of receptors, maximize precision, and guarantee clear, high-fidelity conversations.

DRIVERS WANTED

Ross Harrison was the only biologist to win a Nobel Prize and have nothing to show for it—he was selected by the Nobel committee in 1917, but because of the World War, the prize in physiology or medicine that year was not actually awarded. So let's give Harrison his due. He was, after all, the man who invented tissue culture, and he used it to answer a critical question about the development of the nervous system: How did peripheral nerves, such those innervating the muscles, body organs such as the heart, or sense organs in the skin, connect with their targets?

It was a problem that had also intrigued the master. Cajal's contemporary Viktor Hensen claimed that the neuron and peripheral sense organ began as a single cell that stretched from brain to periphery and then divided. Cajal and a number of other prominent anatomists disagreed. Neuron and target developed independently, they argued, and the axon of the nerve cell grew outward to meet the organ destined to be its partner. The resolution of his silver stain had even allowed Cajal to visualize the uncommitted tips of young axons and to track their progress in serial sections prepared over the course of embryonic development.

Beautiful and detailed as they were, Cajal's slides were only snapshots, however. Harrison's method of culturing living tissue outside the embryo allowed him to actually visualize the movement of axons in real time. To create a micronursery, Harrison dribbled a ring of wax on a microscope slide. He placed a drop of lymphatic fluid on the surface of a glass cover slip—a substitute for the fluid milieu of the embryo. Then he submerged a tiny piece of tissue excised from a frog embryo in the drop and inverted the cover slip over the ring of wax. The wax sealed out microorganisms and prevented dehydration, while the cover slip permitted a clear view of the goings-on within the "hanging drop." Harrison observed that cultured explants of presumptive epidermis or mesoderm—precursors of target organs—made no attempt to explore their new environment. However, when he examined explants from the neural tube in tissue culture, Harrison saw young axons poking out of the explant and growing toward the edges of the drop, as if searching for their missing targets.

But how did they ever find their way? Perhaps they navigated like most men, to whom maps are an impediment and asking directions an insult—they set off in the general direction confident they'll recognize their destination when they get there. Or perhaps evolution put a wife in every axon, an instinct unafraid to consult a map. Cajal favored the latter. The free ending of the axon, he noted, "appeared as a concentration of protoplasm of conical form," a structure he called a "growth cone." He imagined this growth cone to be an inquisitive, groping structure, a sense organ "endowed with amoeboid movements" and possessing "an exquisite chemical sensitivity." All this prying and probing enabled the growth cone to "read" chemical signposts marking the path to the target; as it crawled from marker to marker, the little navigator hauled the axon along behind like a length of cable.

Cajal was correct. Homing axons are not gifted with a wiring diagram, written in the genes—a solution to the routing problem that would have required a supercomputer-sized genome. Nor do they have an internal compass. But they have not been abandoned to wander in the dark either. Every millimeter of the way the young axons can rely on chemical signals to chart their course, direct them around corners, and tell them when to steer clear, steadily and accurately guiding them to their posts.

"Philadelphia—45 miles." In my mind the journey from my house in the suburbs to the University of Pennsylvania, where I'll be using the medical library, can be broken down into several steps. First, there's I-95, leading into the city. Then I have to navigate the expressway through the city, the fearsome left-hand merge onto I-76, and the turn onto South Street. Once I'm in the university neighborhood, I'll head for my favorite parking garage and hope it's not full. Finally, I'll ease the car into a parking space—or bully a student into yielding one on the street.

En route, I'll be relying on dozens of road signs to show me the way. Lines on the pavement will mark out lanes, the median, and the shoulder of the road. Overhead and on either side are signs that tell me "Go here": the green exit signs for the Vine Street Expressway and I-76, street signs for Chestnut Street and Walnut Street, and the white sign at the entrance to the parking garage. And there are signs that warn me *not* to go in certain directions as well, such as "Wrong Way—Do Not Enter," posted to keep people trying to get on the highway from turning onto the off-ramp, and the one-way signs that are the bane of city driving.

Growing axons en route to their targets travel in stages, too, and they'll also be on the lookout for road signs to help them make decisions as they go. Some, like the white and yellow lines that indicate my lane, will define their road. Others will tell them when to turn as well as whether to go right or left, the equivalent of exit or street signs. Still others will block their progress: Do Not Enter signs, or Stop signs planted at their targets to let them know that they should park, fold up their growth cones, and set about building a synapse.

The developing spinal cord is an example of a master cartographer. When we last saw this part of the nervous system, young neurons had scarcely begun to think of the long journey ahead. Just born, they were still sorting out the details of their identity. Now they're finally ready to head out: motor neurons to the muscles; interneurons to the neighbors; and a third group, known as commissural neurons, to partners on the opposite side of the spinal cord. These cells, which will coordinate the left and right sides of the body when they're fully grown, have the hardest route, featuring an especially tricky turn. Neurons starting out on the left, for example, first grow down the wall of the spinal cord toward the floor plate. But before they hit bottom, they'll have to make a 90-degree turn, cross the midline, and continue growing on the right side of the cord, without accidentally wandering back to the left.

Culture a piece of spinal cord with a bit of floor plate and commissural axons grow out of the cord explant and toward the floor plate as if drawn by invisible magnets. The big attraction is actually a

A road map for neural development. Growing axons, like drivers, depend on a variety of road signs to keep them on course as they move toward their destinations. *A*, the adhesion proteins known as laminins, which bind to integrin receptors on axonal growth cones, act as "lane markers" to keep axons on the correct path without straying. *B*, chemoattractants, exemplified by the netrins that direct commissural neurons in the spinal cord to the midline, are the neural equivalent of exit or street signs. *C*, signals like the ephrins post Do Not Enter signs to keep axons away from off-limit areas.

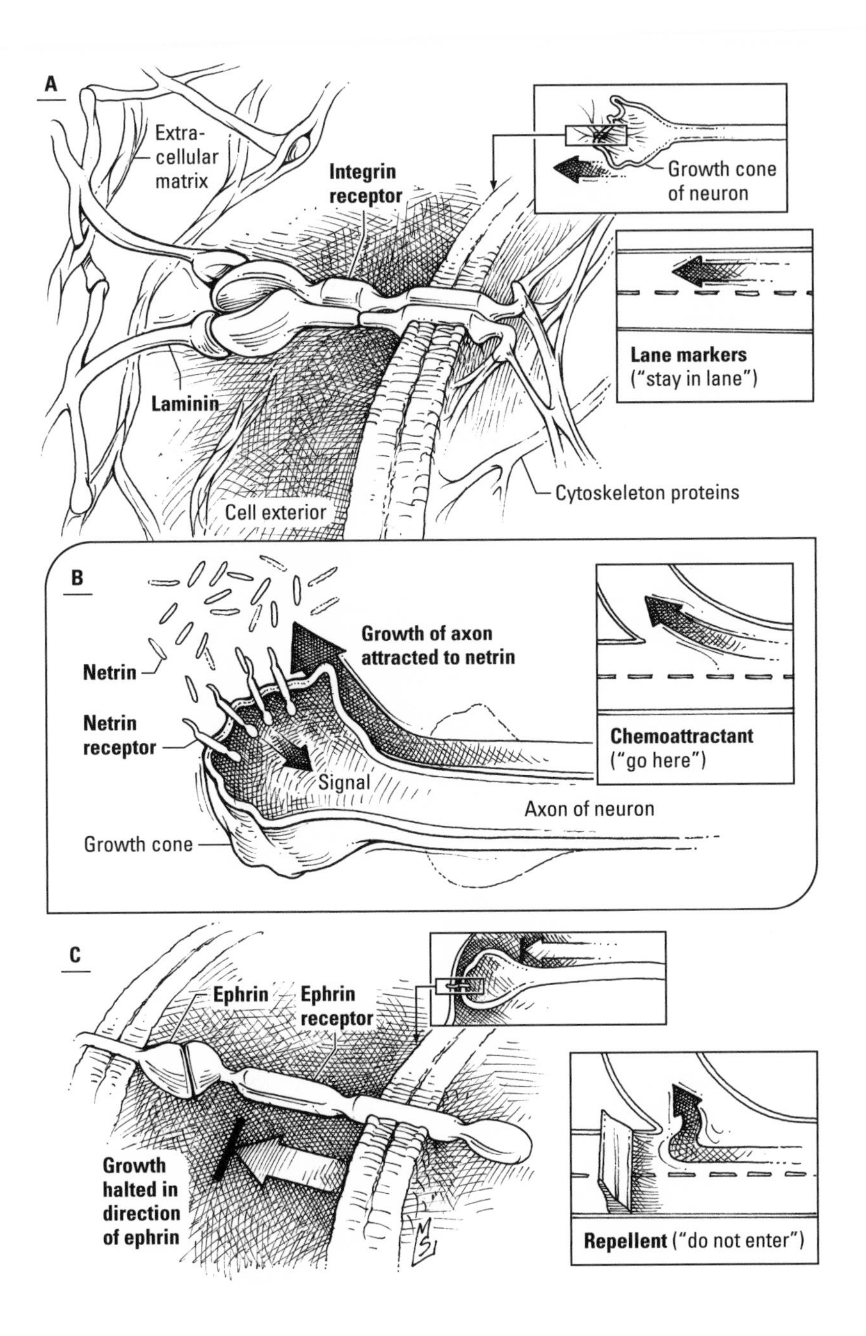
A
Extra-cellular matrix
Integrin receptor
Growth cone of neuron
Lane markers ("stay in lane")
Laminin
Cytoskeleton proteins
Cell exterior
B
Netrin
Growth of axon attracted to netrin
Netrin receptor
Chemoattractant ("go here")
Signal
Axon of neuron
Growth cone
C
Ephrin
Ephrin receptor
Growth halted in direction of ephrin
Repellent ("do not enter")

pair of signaling proteins called "netrins," which direct the growth cone via receptors that mimic the receptor tyrosine kinases favored by growth factors, engaging an adaptor protein studded with SH2 and SH3 domains to activate a GTPase. In the living embryo, netrins are "go here" signals; like highway exit signs, they point commissural axons toward the midline. Netrin-1 is secreted by the floor plate cells. Netrin-2 is spoken by cells just above the floor plate. The ventral-to-dorsal gradient of netrin-2 lures axons into the second gradient of netrin-1, which guides them downward and inward, toward the center of the floor plate.

Anyone who's still confused can flip back a page and consult two markers that ought to be familiar from the previous stage of development—Sonic hedgehog and the bone morphogenetic proteins. Already loitering around the floor plate after it's finished the task of patterning the ventral spinal cord, it's understandable that Sonic hedgehog would take up a new career as a guide to migrating axons, supplementing the efforts of the netrins. Overhead, BMPs have nothing to do either; they might as well busy themselves pushing any errant axons away from the dorsal half of the spinal cord. "Thus, together, these findings provide a pleasing result," write developmental neurobiologists Frédéric Charron, Marc Tessier-Lavigne, and colleagues. "Shh and BMPs, which initially cooperate to pattern cell types along the dorsoventral axis . . . later appear to cooperate to guide commissural axons . . . the attractant is at the ventral midline, providing a pull, and the repellant is at the dorsal midline, providing a push."

Once they reach the midline, whether they're drawn by netrins, pulled by Sonic hedgehog, or shoved southward by BMPs, the axons find a carpet of adhesion molecules rolled out for them across the midline. Receptors on the growth cones allow them to tiptoe from one side of the cord to the other along this runner. Then, before they can change their minds or wander off course, they lose their taste for adhesive proteins. As a result, they cannot go back the way they came; that road's blocked, and they're forced to follow a new path in their new location.

At the other end of the nervous system, you may encounter a clutch of axons crowding the entrance ramp leading from the retina into the brain. If you had been here a few hours earlier, you might have caught a road crew of glial cells, those occupants of the nervous system that didn't make the final cut to be neurons, painting fresh lines to mark the highway from eye to brain. Barred from using white or yellow paint, however, they turned to the adhesive protein laminin, squirting a stripe of this indicator directly into the "skin" of the extracellular matrix that coats their surfaces. Receptors in the growth cones of retinal axons adhere to the laminin in the matrix, much as your finger might stick momentarily to the tacky surface of not-quite-dry paint. Laminin is no ordinary glue, however. This adhesive is also a signal, and laminin receptors, known formally as integrins, relay this signal to the cytoskeleton (as well as to intracellular signal transduction relays).

Look closely at integrins and you'll note an uncanny resemblance to G proteins. A composite of two subunits, α and β, the binding domain of an integrin contains a propeller similar to Gβ, and an I/A domain comparable to Gα. When binding a laminin, the propeller and the I/A domain separate (just as Gα and Gβγ dissociate when stimulated by a G protein–coupled receptor), activating the integrin and dispatching a signal to the cytoskeleton via a short tail that spans the membrane; in the case of the neurons waiting to leave the retina, the message reads: "You're on the right path." The fibrous cytoskeletal proteins tighten, and the growth cone crawls along the laminin-coated surface, translating a chemical signal into movement. Once the axons arrive at their target in the brain—the optic tectum, a way station for visual information—the line of laminin that got them there fades. And the glia don't bother to replace the laminin because the axons wouldn't see it anyway; they've stopped making the integrins that allow them to read it.

The plan then calls for retinal axons to select partners according to their site of origin in the retina, mapping out a faithful representation of the visual field. Axons originating in the temporal portion of the retina (that closest to the side of your head) seek and synapse with neurons in the anterior portion of the tectum, while neurons originating in the nasal portion of the retina (near the bridge of your nose) home to the posterior tectum. To ensure that temporal axons stay on their side of the optic tectum and find anterior partners, posterior tectal neurons have posted "Do Not Enter" signs—signs that can only be read by the temporal axon growth cones. Like Delta (of Notch and Delta fame), these signals, ephrin A2 and A5, are intrinsically well suited to the delicate task of patterning at close quarters, for they remain membrane bound and are never secreted. Feisty ephrins also talk back. After they bind to their receptors, they don't just sit there or fall apart—they post a letter to themselves as well, triggering a signaling relay that travels backwards into the signaling cell.

Contact between a misguided temporal axon and a posterior tectal cell is as toxic as two young children in the back seat of a car. "MOMMM! She touched me!" the axon whines and then retreats to the ephrin-free anterior tectum where its ephrin, or "Eph" receptors can avoid contact with the offending signal. "Ewww, cooties!" retort neurons of the posterior tectum when their ephrin signaling molecules alert them to an out-of-place axon, and they retaliate in some way that scientists haven't quite figured out but suspect may be a change in adhesive properties. The axons of neurons in the nasal portion of the retina, on the other hand, are like good-natured siblings who get along with everyone. Then again, it's easy to overlook insults when you're deaf. Nasal axons have fewer Eph receptors than their temporal counterparts, so they never hear the posterior tectum's ephrin insults. They sail right by the barricade, each on its way to a lifetime partnership.

A blue and white North American moving van pulls up and begins unloading couches, chairs, tables, beds, boxes of dishes and boxes of books, a cedar play set, three televisions, and a plastic dog house. The new neighbors are moving in, and like other experienced suburban home owners, they have arrived with all the baggage and modern appliances needed to get their home up and running. The axon that's just completed its journey and the dendrite it has selected as a partner also pull up to the curb fully prepared to begin the task of setting up a new synapse. Before they even trade a handshake, the axon has already begun to assemble vesicles and is putting the final touches on its secretory apparatus, while the dendrite already has begun to stock up on receptors. Both are dying to strike up a conversation. But the words they'll trade now will not be anything like the gene-manipulating signals that made them neurons in the first place.

The signals that control the formation of synapses, says Joshua Sanes, a neurobiologist at Washington University School of Medicine in St. Louis, "are organizers, not inducers." He explains: "Inducers make the cell into something it didn't used to be by turning on a lot of genes. During synapse formation, proteins that neurons have already made—the docking mechanism for vesicles or calcium channels, for example—are being put together." What's more, Sanes notes, the assembly of synaptic proteins doesn't really require formal neurotransmission, although the two neurons certainly talk to each other during the process. Once the two neurons pair off, the presynaptic neuron will install the secretory apparatus it has constructed, stock it with a supply of vesicles, and begin releasing little trial balloons of neurotransmitter. The postsynaptic neuron will follow its example, gathering receptors into clusters and fixing them in place with the protein matrix of the postsynaptic density. Finally, adhesion molecules will bind the whole structure together, securing the connection and sealing it off to keep neurotransmitters in and distractions out.

Because the connections between nerves and muscles are easier to observe and manipulate during development, researchers who study synapse formation, like Joshua Sanes, have come to rely on the neuromuscular junction as a model system. In this case, the presynaptic element is the motor neuron, which will use acetylcholine as its neurotransmitter, and has a branching axon looking to form synapses with dozens or even hundreds of different muscle fibers. The postsynaptic element is the immature muscle fiber, or myotube. As the synapse matures, the surface of the muscle will wrinkle into a series of so-called junctional folds; the acetylcholine receptors cluster at the tops of these folds. Between axon and muscle, in contrast to neuron-neuron synapses, is a thin band of matrix proteins, the basal lamina, which runs through the middle of the synaptic cleft like a line drawn to emphasize the separation of the two elements.

The formation of the neuromuscular junction is initiated by the arrival of a motor neuron axon eager to latch on to a myotube and take charge of its life. Young myotubes have already begun making acetylcholine receptors. But without someone to explain how to gather the receptors into clusters, each has simply inserted them randomly into its membrane. With a word—agrin, a large protein encrusted with carbohydrate residues—the axon herds those receptors into the synapse and secures them with a cytoskeletal protein called rapsyn. Motor axons like their receptor clusters to be really generous, so in addition to agrin they release a signal, neuregulin, that binds to a receptor tyrosine kinase on the muscle fiber and calls for more acetylcholine receptors to fill out each cluster. Finally, when the big day arrives and the motor neuron starts firing action potentials in earnest, the acetylcholine it lobs at the muscle fibers shuts down any illicit receptor making or gathering.

The muscle fiber and the intervening basal lamina have had a few choice words for the axon as well. Some scientists say they've overheard cadherins, similar to those that Velcro neuron to neuron at central nervous system synapses, telling motor axon terminals to ramp up the supply of synaptic vesicles and reinforce the protein scaffolding. Laminin, a constituent of (surprise, surprise) the basal lamina, is suspected of exhorting them to grow up and start releasing neurotransmitter. And to get axons in tip-top shape for neurotransmission, some recent studies suggest that muscles feed them a protein shake containing growth and survival factors.

Synapses in the central nervous system are made in much the same way, says Sanes, although far less is known about the identity of the organizational signals. For example, when immature neurons are maintained in tissue culture, axon terminals stimulate the clustering of receptors for glutamate and the inhibitory neurotransmitter GABA (gamma-aminobutyric acid) in much the same way motor axons stimulate the clustering of acetylcholine receptors in muscle fibers. In the brain, however, many neurons must also perform skilled feats of receptor distribution, for their dendrites may make synapses with numerous axons, each of which uses a different neurotransmitter. Scientists don't know what these axons say to engineer such complex receptor patterns. But they do know that neurons in the central nervous system have a protein on the postsynaptic side much like the rapsyn in myotubes, which helps stabilize the clusters once the mystery signals collect them. Conversely, on the presynaptic side, a recent study has demonstrated that the adhesion molecule neuroligin tells axon terminals it's time to assemble synaptic vesicles.

When neuronal migration and synapse formation are done, the young brain can feel a real sense of accomplishment—it's wired about 10^{15} individual connections. If anything, it has done its job a little too well. While obsessing about precision, it overlooked capacity; busy planning itineraries, it has overbooked most destinations. Many muscle fibers, for example, have been innervated by not just one axon, but several. The nervous system that has had too much of everything all along—precursors, neurons—now has too many synapses as well. The extras must be eliminated, all but one axon withdrawn. That, says Joshua Sanes, marks the beginning of a lifetime of adaptation in which alterations to synapses play a pivotal role. "The initial contacts are the hardwiring," he explains. "Synapse elimination is the refinement. And it probably happens in the central nervous system as well. We're used to thinking of this incredible specificity. The final pattern does come partly from hardwiring, one on one, but also partly by rearrangement."

A surfeit of synapses may be a fail-safe, a way of ensuring no part of the wiring pattern is accidentally left out—"like shooting at a target with a machine gun, rather than a pistol," as Sanes puts it. Or it may be a matter of survival of the fittest: from the pool of synapses the brain selects the "best" ones. But there's also a third explanation, Sanes adds—the extra synapses offer a way to sculpt the young brain according to its experiences. "The best way to make patterns may be to selectively eliminate the wrong possibilities. In other words, you could build only the proper synapses. Or you could build extras, stabilize the proper ones, and let the others fall away."

Neural synapse formation can take place in the complete absence of neural activity. Synapse selection and maintenance cannot. "Activity is essential for growth and maturation," concludes Sanes. "Experience is activity, and so experience is what shapes synapses and our nervous systems."

SPEAK, MEMORY

A whisper of leaves, a sudden shadow, an agile duck-and-scramble under a log. A quick response has just averted certain death in the clutches of a hungry predator, but what good is such an accomplishment in the long run unless the experience can be retained for future use? A quarter-mile away my dogs and I wander along the creek bank, enjoying a languid summer afternoon. We don't have to worry about predators. But this daydream could become an ordeal if I don't recognize the poison ivy growing along the path or can't find my way back to my car.

An organism's ability to mount effective, appropriate, and adaptive responses depends on the brain's skill at differentiating the essential from the irrelevant and applying lessons learned in the past to the present. Sifting through what is seen, heard, felt, read, or imagined through the filter of experience, it must correctly identify a bewildering variety of people, places, things, and events and assign each a meaning. That rustle means "Danger! Run!" A trio of shiny leaves means "avoid this plant." The bend in the creek and a map in my head tell me how to retrace my steps to the parking lot.

The brain isn't born with a recording of a fox's footfall, the knowledge that certain plants irritate the skin, or a map of the nearest park. As Steven Hyman, Harvard provost and former director of the National Institute of Mental Health has put it, "Organisms need to be able to make predictions about where danger lies, where good, life-enhancing things lie. Some things—like the rewarding nature of sex—are hard wired. But there are a lot of things that evolution could not know about ahead of time."

Behavior evolved to facilitate change. But the nervous system that coordinates behavior also includes mechanisms to facilitate continuity, to enable considered responses as well as allowing for novel ones. Taking advantage of the ability to link the binding of chemical signals to changes in proteins and gene expression afforded by the syntax of cellular language, neurons have exploited synaptic transmission to devise methods of generating connections between behaviors and outcomes—learning—and to store important associations for future reference—memory.

No doubt about it, the tree had to come down. It had passed the point of simple encroachment and annexed an entire corner of the yard, adding a swath of the deck for good measure. It swallowed the sun and regurgitated a mound of sickly yellow needles in its stead. The new neighbors contemplated the behemoth for two weeks, then called a landscaping service to cut it down.

Out with the old and in with the new—swings, the sandbox, the wading pool. Grass sprouted. Weeds invaded. Even the climbing roses on our side of the fence, anesthetized for so long by the shadow of pine boughs, woke up and turned somersaults over the fence rail, tossing sprays of blooms as they went.

During development and throughout life, experience, that tireless gardener, tends the synapses of the brain, taking away a little here, adding a little there. "Use it or lose it" is the order of the day. Neglected, underutilized connections grow weak and are pruned. Active synapses grow stronger. Their participation in a clever association or a beneficial response is noted and encouraged. Used and reused, they sprout a memory—literally. Experiences this significant cannot be entrusted to some transient medium; to preserve them for months or years to come, they are recorded in the actual physical structure of the neuron.

The climbing roses here are the dendrites—or to be more specific, the knobs known as spines. Shaped like tiny clubs, with a bulbous head seated on top of a narrow shaft, spines are hot spots for synapses, particularly so-called excitatory synapses, in which the word spoken by the axon terminal stimulates the postsynaptic cell to action. The precursors of spines are mobile, actin-rich fingers, or filopodia, that grasp at incoming axon terminals in the earliest stages of synapse formation. But spines retain their penchant for shape shifting long after the building is over, making them a malleable medium, easily sculpted or redistributed.

"Action!" at an excitatory synapse, as noted earlier, is translated "glutamate." Released by the presynaptic partner, it can choose between two types of glutamate receptor across the cleft, both signal-gated ion channels fashioned specifically for the speedy propagation of important messages. AMPA receptors conduct sodium and potassium ions and are incorrigible chatterboxes. NMDA receptors conduct calcium as well as sodium and potassium. Strong, silent types, they can't talk much of the time even if they want to because their mouths are usually plugged with a fat magnesium ion. Only a combination of glutamate and a conversation animated enough to activate, or depolarize, the postsynaptic cell will expel the magnesium, freeing the NMDA receptor to talk calcium to listeners inside the cell.

Cyclic AMP has the glamorous history, but no second messenger is more hard working or widespread than calcium. Within the cell, changes in calcium concentration triggered by events like the opening of NMDA receptor channels are detected and relayed to other signaling proteins by dedicated calcium-binding proteins. The most ubiquitous of these go-betweens, found in all eukaryotic cells, is called calmodulin. Calcium turns calmodulin into a social butterfly. Content to stretch out languidly to its full length when calcium isn't around, calmodulin grabs the nearest target protein in a bear hug when calcium is present; among the favorite objects of its affection are a family of kinases that add phosphate groups to the amino acids serine or threonine, known collectively as "calcium-calmodulin-dependent protein kinases"—CAM kinases for short. An active CAM kinase is memory's best friend. Thanks to its eagerness to gossip with other signaling proteins, neuronal activity can be translated into long-lasting changes in the structure and function of spines and synapses.

Even a slug can learn. With a nervous system of only 20,000 neurons, the giant marine snail, *Aplysia californica*, can be trained to respond to an innocuous stimulus as if it was a life-threatening event. Following a brief electric shock to its tail, the slightest touch to the overlying mantle or the siphon causes the animal to flinch and retract its gill. Neurobiologists call this exaggerated defensive reaction "sensitization," and its duration varies according to the intensity of the noxious stimulus. The effect of a single shock lasts only a few minutes. A series of shocks, however, elicits vigorous reactions to touch for days or even weeks.

Aplysia isn't a rocket scientist, but for neurobiologist Eric Kandel, corecipient of the 2000 Nobel Prize in Medicine or Physiology for his work on the biological basis of memory, that was the appeal of its simple accomplishments. When Kandel began studying the neurobiology of learning and memory in the 1970s, trying to approach the problem by studying the sophisticated mammalian brain was as impossible as searching for the proverbial needle in a haystack. What was needed instead, he recalls, was a "radically reductionist approach" that focused on the acquisition of a simple task in a creature with a less intricate nervous system. By doing so, Kandel believed he could identify fundamental neural mechanisms common to memory storage in all nervous systems. "When I went to *Aplysia* I had a sense that there was a universality to biological systems," he observes. Indeed, his studies of simple learning paradigms revealed two basic features of memory common to both this simple invertebrate and creatures like mammals, capable of mastering more difficult tasks: first, that all forms of learning, whether they involve simple associations or the retention of facts and events, occur in two phases, one transient, or short-term, and the other more enduring; second, that short-term memory requires only modifications to existing proteins, while long-term memory storage entails the synthesis of new proteins and modifications to the physical architecture of the synapse.

The synaptic changes associated with short-term sensitization to an electric shock are a metazoan adaptation of an idea known even to bacteria: a reminder can be as simple as a Post-it note slapped on a protein. To "remember" the concentration of a nutrient from one twiddle or tumble to the next, *E. coli* adds or subtracts methyl groups to and from its chemotaxis receptors. In the slug, news of an electric shock to the tail is transmitted from a sensory neuron in the tail to an excitatory interneuron, which in turn synapses on a sensory neuron in the siphon and regulates its activity. The exchange between the interneuron and the siphon sensory neuron results in the activation of protein kinase A. The kinase, in turn, phosphorylates calcium channels and proteins of the secretory apparatus, enabling the sensory neuron to release more neurotransmitter—the alarm signal that activates motor neurons responsible for gill withdrawal.

Post-it notes are easily lost. Important information—like repeated electric shocks—merits a more enduring sort of record keeping. In this case, recurrent activation of the interneuron leads to persistent activation of protein kinase A. It's so annoyed, in fact, it complains to the MAP kinase. "You're right," MAP kinase agrees, "the higher-ups need to know about this," and so it scuttles off to the nucleus and pours its heart out to the transcription factor CREB. CREB talks to some genes, the genes take action, and the first thing you know, new proteins are on their way to the sensory neuron terminal, proteins that will physically reinforce this obviously important connection by stimulating the addition of new synapses.

So what if a sea slug can learn to avoid live wires, you may be thinking. A human being needs to remember more important things—the names of co-workers, how to log on to the new office computer system, how to care for a newborn baby, the dates and battles on tomorrow's history test, lists of signaling proteins. This sort of memory, the one we associate with "book learning" or factual information, is called "explicit memory," and for that you need a part of the brain known as the hippocampus.

The name means "sea horse"—that's more or less what this structure looks like, curving along the inner wall of the temporal lobe, that part of the brain located just above your ear. Central to the neural circuitry responsible for explicit memory, the mammalian hippocampus, as Kandel had observed in his pre-*Aplysia* days, was an impenetrable thicket of crisscrossing fibers that defied researchers eager to infiltrate its archives. Then, in 1973, Timothy Bliss and Terje Lømo discovered that applying a brief train of high-frequency electrical stimuli to neural pathways in the hippocampus could evoke a striking activation of postsynaptic neurons, a phenomenon they called "long-term potentiation," or LTP. Readily studied in the living animal as well as in laboratory preparations, LTP made the hippocampus accessible and proved a powerful model for the study of explicit memory processes in the mammalian brain.

Like sensitization in *Aplysia*, Kandel and others found that LTP has two phases, an early phase and a late phase. Being excitatory neurons, the terminals of cells that make up LTP-sensitive pathways broadcast glutamate, which binds to AMPA and NMDA glutamate receptors on dendritic spines. Everyday low-frequency conversations engage only AMPA receptors, with no lingering aftereffects. But the high-frequency stimulation used in LTP triggers powerful depolarizations along with the release of glutamate. The NMDA receptor sneezes, the magnesium stuck in its channel flies out, and calcium flows into the dendrite. Once inside, it activates calcium-sensitive kinases. The kinases, in turn, phosphorylate AMPA receptors, increasing *their* sensitivity. In addition, the fuss draws more AMPA receptors to the synapse and encourages the expansion of the postsynaptic density. The end result is a stronger synapse, an effect that lasts two to three hours.

Once again, however, memories based on such cosmetic changes to synaptic proteins fade like a fresh coat of paint. Remembering facts and figures takes practice; in the laboratory neurobiologists replicate the beneficial effects of studying by applying several successive trains of high-frequency stimuli. Now calcium has a more interesting tale to tell, and it searches out a new confidant: adenylyl cyclase. News of the tryst reaches MAP kinase by way of that little tattletale, cyclic AMP, the kinase hurries into the nucleus to tell CREB, and CREB spreads the word to genes. Soon a construction crew of new proteins is on its way to the dendrites, ready to build a monument to high-frequency stimulation that will last at least 24 hours.

First, however, the crew has to find the construction site. "When you turn on genes, the products are sent to all of the synapses. But each neuron may make as many as 1,000 connections, involving more than 100 different cells. So the question is, are you enacting a cellwide change or a synapse-specific change?" Kandel observes. And he found that the answer lies in some unusual furnishings of dendrites: mRNA transcripts, shipped and stored in an inactive form prior to all the excitement, and the machinery to carry out local protein synthesis. All the synapses have these materials. But only those "branded" during the short-term memory phase will now be allowed to use them. "You have a local marking signal," says Kandel, "and the function of this is that it allows you to use the proteins made when genes are activated at that one synapse. mRNAs, shipped in dormant mode and activated locally, give rise to the growth of synapses."

In effect, Kandel concludes, "neurons have two genomes. They have the genome that's in the nucleus and then the machinery for expressing genes locally, a duplication essential to their ability to learn and adapt."

People say that horses are "not smart" or that they are "flight animals." But I think they are actually a lot like teenage girls. They like to run in herds. They pick on outsiders. They question authority figures. They think everyone and everything is out to get them. And when they're in trouble, they can be counted on to panic.

Then again, inexperienced riders can act a lot like parents. They lecture instead of listening. They pay more attention to experts than their own intuition. They're better at making rules than enforcing them. And they're really, really bad at dealing with panic.

"Because I said so, that's why!" My horse and I were locked in the same old argument about her head tossing; exasperated, confused, and shamed by advice from the more experienced to "stop letting her take advantage of me," I had gone parental, resorting to an attitude already proven to be unsuccessful for dealing with adolescent infractions. She was returning the favor by acting juvenile: "I don't get it! You don't understand what it's like to be a horse! You can't make me!" Conditions were perfect for a panic attack. On a calm, windless day, in a part of the world where a predator hadn't been sighted in 50 years, carrying out a routine maneuver, she shot sideways so violently my neck snapped and began running.

The road was only a few hundred feet away and I knew—cold, sinking, panic knew—we were going to end up right in the middle of it. With my only other option a collision with a rapidly moving vehicle, I braced my hand on her neck and tugged repeatedly on one rein until I finally hauled her to a spinning stop. She bucked and I bailed. I walked away from what might have been a train wreck with nothing more than muddy clothes and a sore neck. But two days later I got back on my horse and I was afraid. A week later I was afraid. A month later I was afraid. And the fear wouldn't go away.

A brain that can remember the birth of a child, how to play the piano, multiplication tables, the names of dozens of business associates, and the rules for filing an expense account can also learn to be afraid—and remember this lesson even longer than a sea slug remembers a series of electric shocks. Like any other high-intensity stimulus, terrifying events can remodel the brain, re-creating the traumatic event in the structure of neurons. The heavy lifting in this emotional memory is not the hippocampus, however, but an almond-shaped assemblage of cells about as big as the tip of your thumb, perched on the head of the hippocampus like a hood ornament. Known as the "amygdala" (meaning "almond," after its shape), it is "the anatomical equivalent of city hall," a crossroad where inputs from brain regions that traffic in sensory data, the perception of bodily sensations (like the bottom-dropping-out-of-your-stomach feeling you might get when racing toward a busy road on an out-of-control, thousand-pound animal), associations, rational thought, and other older memories (courtesy of the hippocampus) intersect. In other words, the amygdala is perfectly located and hardwired to connect the details of an event and the feelings they inspire, and it's no surprise this structure has been linked to emotional behavior and emotional memory for more than a century.

In the laboratory, neuroscientists induce the pairing of neutral stimuli and fearful emotions using a paradigm that's a combination of Eric Kandel's technique for sensitizing snails to touch and Pavlov. A rat is placed in a cage with a floor that is actually an electrified wire grill. The animal's first experience in the cage is uneventful—the only disturbance a mysterious beep coming from a speaker set into the wall of the cage. During the next visit, however, the sound is followed each time by a brief, harmless but unpleasant shock. Just a few shocks and the sound alone is enough to make the rat freeze in place, its heart pounding and its blood pressure soaring. The memories implanted by this "fear conditioning" are extremely long lived; months later a single reminder—in fact, any kind of stress—can reawaken it.

Joseph Le Doux, an expert on the neurobiology of emotional memory, and others have demonstrated that fear conditioning induces long-term potentiation in the amygdala much as high-frequency electrical stimulation induces LTP in the hippocampus. The development of ultrasensitive neural responses correlated with the emergence of the fearful response and, much like the animal's anxiety at the sound of the tone, the sensitivity of amygdala neurons persisted long after the training period. And only rats that learned to be fearful developed LTP-like responses; no changes in neural sensitivity occurred in the brains of rats exposed to the tone but never shocked.

These observations suggest that fear persists after a traumatic experience because the chemical fireworks ignited in the neuronal pathways that converge on the amygdala set in motion the same processes—the phosphorylation of receptors, the inward flow of calcium, the generation of new synapses—that underlie learning and memory in the simple circuitry of *Aplysia*. Such restructuring in the wake of events important enough to elicit animated conversations between neurons evolved to promote survival, to help us recognize potentially lethal situations and take action before it's too late. But when fear endures long after the original danger is gone, it can become debilitating instead of advantageous.

"If you want a building to learn," writes Stewart Brand, "you have to pay its tuition." The price we pay for having an adaptive nervous system in which memory is possible is that memory will be too effective, that outdated, inaccurate, or misleading information will be retained along with the good and the useful. Welded to the facade of proteins or built into the membranes of dendrites, such misinformation cannot be erased quickly or easily.

Denise is another example of the grief that can ensue when neurons retain the wrong things. "I always said that if they ever came to take away my children, that would be the end of the world. I would stop," she says. Stop abusing cocaine, she means. Too bad the experience of getting high had already written itself into her brain.

"For a while, I still used to go to work every day, cook dinner, take the kids to the mall on Friday. Then I stopped going to work. I stopped paying the bills. The car was gone, next the house was gone. And then the kids."

"I had two at home with me, and they took them while those kids were crying and holding on to the banisters. Then they went up to the school. My daughter called me and said, 'Mom, I'm scared. Someone's here to take me away.' It hurt me more than anything ever hurt me in my life. And still I couldn't stop."

All substance abusers take drugs. Those who become addicted, however, are changed in a fundamental physical way by the experience. As a result, they can't stop taking drugs, even in the face of profoundly negative consequences.

Their predicament stems from the fact that cells communicate in chemicals, meaning that cellular communications, including discussions between neurons, are accessible to outside influence. A drug that looks like a neurotransmitter, socializes with its receptors, or interferes with its clearance once transmission is completed can trigger the same sequence of events as a legitimate synaptic discourse—including alterations in gene expression. As a result, neurons can "learn" from their experiences with drugs just as they learn from other life experiences and retain this information for long periods of time.

Learning from drugs can be helpful. Antidepressant medications like Prozac and Zoloft, or the antipsychotic medications used to treat schizophrenia, for example, use their influence to calm troubled neurons. The adaptations they induce relieve symptoms, restoring patients' ability to lead productive lives.

Addictive drugs tell beautiful lies. They also remodel the brain, but the adaptations they induce are anything but beneficial. Over time, the small changes in gene expression that follow in the wake of every exposure to the drug accumulate and lead to profound and persistent changes in the structure and function of neurons. When these changes reach a critical threshold, drug abuse slides over into addiction and drugs are no longer an option but a compulsion.

Drug-induced adaptations are concentrated in the neural circuitry governing emotion and motivation—particularly so-called reward pathways. Here, their favorite targets are the neurons that use the catecholamine dopamine as a neurotransmitter. Under normal circumstances the wavelet of dopamine these neurons emit when they want to announce a positive experience—a delicious new food, for example, or a behavior that led to praise—records the event along with the label "good," increasing the likelihood the behavior will be repeated. Addictive drugs add bigger labels and flood the synapse with "levels of dopamine never found in nature," according to neurobiologist Eric Nestler. The surge of dopamine masquerades as an urgent message: "This is the most important thing that's ever happened to this brain—and don't forget it!" Too much exposure to this noise and drugs come to seem more essential than anything else—even your children.

A nervous system built to adapt gives us the flexibility to cope with a wide range of contingencies. But when traumatic experiences or addictive drugs muscle into the signaling pathways at the heart of this flexibility, the brain can also be waylaid by its ability to change. Or in Steven Hyman's words, "Not all adaptations are good for the organism."

Memories engraved in the fabric of the brain cannot be willed away, ignored, or rationalized. "If you can't get over your fear factor, maybe you should quit riding," one self-styled expert advised in response to my unrelenting anxiety. It's an attitude someone like Denise would recognize all too well: "If you can't give up drugs, you should quit raising children." Unfortunately the brain just doesn't work that way. The moral imperative of neurons is to ensure the survival of the body, preserving the memory of experiences that threaten or enhance life regardless of whether they conflict with social standards, circumvent the rule of law, or prove inconvenient at a later time.

The only antidote to destructive memories is constructive memories. I couldn't erase my fear by sheer willpower. But I could learn a different style of riding, better ways to cope with an emergency, and how to spot the little cues that meant my horse was reaching the limits of her patience. And gradually, memories of good rides—a jaunt through the woods on a blue-sky autumn morning, jogging a perfect serpentine—are overwriting the memory of the accident.

I don't know what's happened to Denise since she left the halfway house where I met her. I hope she's at home tonight, cooking dinner for her kids, rather than in the hospital, on the street, or in prison, knowing that like the fearful, the addicted can learn to outwit the memory of drugs. Relapse remains an ever-present risk. But in the long run, teaching people to change is more likely to succeed than putting them in jail, for no punishment, however harsh, can reverse or negate the damage addicts have done to their own brains.

The good news for us, with our adaptable brains, is that flexibility is the agent of hope. As long as neurons converse and synapses multiply, the capacity for change will not diminish. What we learn, we may not be able to unlearn, but we can relearn, replacing memories that have outlived their usefulness with new and better ones.

THE ART OF WAR

The phone rings as you pause to hand your secretary a draft of the new proposal en route to your 11 o'clock meeting. "Wait, it's for you," she says, sneezes, then hands you the phone. You're thirsty after the conversation, so you stop for a quick drink at the water fountain—the same fountain where the receptionist rinsed out six coffee mugs a few minutes earlier. You drop your notepad on the conference table and decide to sample the doughnuts laid out on a table at the back of the room. Good, there's still a chocolate one left. It's the one your boss picked up, then, remembering her diet, put back before you arrived. You sit down, pull yourself up to a table touched by three other people earlier this morning, and rest your hands on the arms of a chair used by two of them as well as someone at last evening's after-hours brainstorming session. As you're shaking hands with the prospective vendor (and the four staff members he's brought along), you realize that you've forgotten to bring a pen. Your neighbor has an extra; fortunately, he's looking the other way when you nibble absentmindedly at the tip. When the presentation is finally over, you shake hands with everyone again, help your pen-sharing colleague collect the used napkins and empty cups, and sprint for the company cafeteria. You think of stopping to wash your hands, but you have only 20 minutes to grab lunch, check on your secretary's progress, and hustle over to the next building for your one o'clock meeting.

Even if you're single, celibate, and profoundly antisocial, you can still have an astonishing amount of intimate contact with your fellow human beings. If this had been an actual workday, you would have inadvertently rubbed shoulders with more than a dozen people. You would not have shared a single detail of your private life—home phone number, shirt size, political affiliation, or the incident that precipitated your divorce—with any of them, but you would have shared something more personal and potentially more damning: their microorganisms. Given the number of bacteria and viruses you could have inhaled, picked up, or swallowed along with your chicken salad sandwich, you're lucky that under the skin you're defended by one of the most effective fighting forces on the planet—the human immune system.

Damn! You're having an especially bad day. You just cut yourself opening a box, and now barbarians—bacteria in the air and on the skin—are swarming over the walls. But don't panic. They won't get very far before scouting parties assigned to detect and report such catastrophes discover them and broadcast a call to arms that will bring a squadron of defenders running to the scene. Alerted to the breech by the telltale signs of injury—the detritus of dead and dying cells, cries of pain clamoring for the attention of sensory neurons—one group, known as mast cells, disgorge a cocktail of cytokines, enzymes, and small-molecule signals like histamine and prostaglandins, synonyms for "help." Others, called macrophages, are more aggressive. Rapacious as lions, they hunt down the invaders, snare them with receptors keyed to surface features common in pathogenic bacteria (in particular, the unique carbohydrates bacteria use to decorate their membranes and cell walls), and devour them alive. In addition, the macrophages collaborate with platelets, fibroblasts, ketatinocytes, and endothelial cells to issue a call for reinforcements. Known collectively as chemokines, these "come here" signals bind to G protein–coupled receptors on other macrophages as well as neutrophils, carnivorous white blood cells packing an arsenal of toxic chemicals that enable them to poison any invaders they don't swallow. Following the sound of chemokines, the new recruits, collectively referred to as "phagocytes," ride the bloodstream to the battlefield.

Before they can join in the fight, however, they have to get back out of the bloodstream and into the infected tissue. Fortunately, the endothelial cells that line the blood vessel are eager to give a soldier a hand. Alerted to the crisis by the ruckus coming from the wound, they grab at the phagocytes with fingers made of selectins, adhesive molecules that bind to integrin receptors. And felicitously, the chemokines that summoned the phagocytes have also unclenched selectin-hungry integrins on *their* surfaces. Grasping and clawing, they roll to a stop, squeeze into the infinitesimal space between endothelial cells, and crawl through this wormhole to the other side. Then they descend on the invaders like the Mongol hordes.

You cannot see this battle, but within hours, you will be well aware of its occurrence. Those calls to sensory neurons will register as pain. The area surrounding the cut will be warm and red—the handiwork of mast-cell missives that order blood vessels to dilate—and swollen as well thanks to signals that weaken the connections between endothelial cells to facilitate the migration of phagocytes, allowing fluid to leak into the tissue in the process. Immunologists call these artifacts of battle "inflammation," the cells that engineer it, the innate immune system. Honed by hundreds of millions of years of evolution, protecting invertebrates and vertebrates alike, this first line of defense is swift and so effective it quells many infections in the earliest stages. But it has one overriding disadvantage: inflexibility. The pathogen-recognition receptors of the innate immune system are hardwired into the genome, a storage facility where space is limited and change comes slowly. Forced to make economical choices, therefore, evolution has compiled a receptor repertoire limited to proteins that correspond to a handful of features common to the majority of microorganisms, rather than a comprehensive repertoire with a unique receptor for every troublemaker. But while animal cells were conserving, their attackers were inventing. Bacteria, replicating as often as every 20 minutes, kept merrily mutating and evolving ways to deceive, evade, or hide from our phagocytes and their hidebound old receptors; viruses simply took refuge inside the cells they preyed on. To keep up with the enemy, animals needed an addendum to the innate immune system that was more like a nervous system. It had to be flexible enough to stay current with pathogens that change their proteins almost as often as a teenage girl changes clothes, yet precise enough to focus the attack on a single microbe.

About the same time evolution came up with the idea of a backbone, it happened upon a solution to the defense problem as well: the RAG (for *r*ecombination *a*ctivating *g*enes) proteins. RAG proteins form an alliance called a recombinase, an extraordinary superenzyme that gives a cell the power to cut and paste its own genes instead of waiting around for evolution to do it. Of course, if all genes could change their organization willy-nilly, mayhem would ensue, so the right to an extreme makeover by the RAG complex was limited to the privileged few that encode certain proteins known as immunoglobulins and one near relative. Rearrangement by RAG added a personal touch to these proteins—every cell that made them could come up with its own one-of-a-kind molecule. Implanted in that cell's plasma membrane, such a protein made a perfect receptor. Make the cell a white blood cell, that receptor, a homing device trained on a single protein fold, loop, or crevice of a pathogen and you have a guided missile. Encourage that cell to divide, and you have a potential army.

This new line of defense, adaptive immunity, extends the range of the immune system by several orders of magnitude. Every phagocyte displays the same battery of receptors, but each lymphocyte (immunologists' name for the white blood cells of the adaptive immune system) is an individual; together, these antigen receptors can distinguish millions of bacterial and viral features. No matter how many foreign proteins an organism encounters in its lifetime, therefore, it's likely to have at least one lymphocyte able to recognize that antigen. "Pathogens are always evolving, but we've evolved, too," says immunologist Ralph Steinman. "The RAG genes can make a receptor for everything."

Lymphocytes come in two flavors. B cells make immunoglobulins called antibodies, which double as receptors and weapons. Membrane-bound antibody detects pathogens, while secreted antibody molecules adhere to them, blocking their access to cells and making them more conspicuous to phagocytes. The antigen receptors of T cells are similar to antibodies, but they are not immunoglobulins. Embedded, never secreted, they resemble other transmembrane receptors, with an external binding site and an internal segment on friendly terms with a gaggle of kinases and other signaling proteins. To their fellow lymphocytes, T cells are helpers that activate B cells and stimulate them to produce antibodies. To macrophages, T cells are generals, directing and deploying them on the battlefield. To bacteria, viruses, and parasites, T cells are cold-blooded killers. No use trying to hide by holing up in another cell; these defenders won't hesitate to murder one of their own just to eliminate a pathogen that's set up housekeeping inside—or call in the macrophages to eat it alive.

Like other blood cells, lymphocytes, both B and T varieties, are born in the bone marrow. But while young B cells are home-schooled along with youngsters studying to be erythrocytes or macrophages, future T cells are sent to boarding school in the thymus. Whether educated at home or abroad, most will not live beyond school-age, for all are subjected to a test to determine if the receptors they make bind proteins made by body tissues; to prevent inadvertent self-destruction, those that fail are encouraged to commit suicide as soon as possible. Survivors complete their studies and are discharged to the bloodstream, checking in periodically with one of the lymph nodes to peruse the antigens on display there. While in this naïve state, they keep their weapons—secreted antibodies, toxic proteases, cytokines—in the deep freeze. If a lymphocyte should actually encounter its corresponding antigen, however, the phrase formed by the interaction of antigenic signal and receptor precipitates the formation of a sentence that orders the lymphocyte to do two things: (1) divide, generating a clone army fixated on the invaders, and (2) arm itself for battle. B cells begin producing and secreting antibodies, while T cells manufacture cytokines (to talk to B cells and macrophages), toxic chemicals (to kill infected cells), and suicide signals (to induce them to commit apoptosis).

Should the bacteria multiplying in that cut on your finger gain the upper hand or cold viruses on the pen you chewed establish a beachhead in your respiratory system, their rising numbers will generate a steady stream of refuse—membrane fragments, half-digested proteins, secretions and toxins—collected and put on display in the lymph nodes. Naïve lymphocytes comb through this treasure trove like bargain hunters at a yard sale; with this kind of selection, it's almost inevitable some will find the antigen of their dreams. Their success is the body's saving grace—within a few days of the discovery, a regiment of active, heavily armed lymphocytes will be speeding to relieve the beleaguered innate immune system. Carpet bombing the invaders with antibodies, confronting them in hand-to-hand combat, and overwhelming them with additional reinforcements, the adaptive immune forces seek to end the enemy's quest for total domination once and for all.

Agile, discerning, and ruthless, the adaptive immune system gives vertebrates one more disease-fighting advantage over creatures armed only with innate immune defenses: once it's fought a particular pathogen, it remembers the experience. After the battle's over and the war is won, most of the lymphocytes engaged in the conflict die, sparing the body the aggravation of containing an unemployed fighting force still spoiling for a fight. A few, however, survive to become cellular war memorials, the image of the defeated enemy preserved in their antigen receptors. Should the same pathogen mount a second invasion, the adaptive immune system doesn't have to build an army from scratch the way it did the first time around—these memory cells spearhead a counterattack that's faster, fiercer, and even more specific than the first. Much as synaptic memory shapes the nervous system to reflect individual experiences, immunological memory tailors each individual's immune system to confront the pathogens endemic to its particular environment, allowing long-lived organisms to withstand repeated encounters with disease-causing bacteria and viruses.

Patrolling the site of an incipient infection along with the phagocytes and straddling the gap between adaptive and innate immune systems is yet another representative of the immune system. In contrast to the stout, rather homely lymphocyte, these cells are resplendent with spines and tentacles that reminded immunologists of the dendritic arbor of neurons, hence their name: dendritic cells. Immature dendritic cells devote their youthful energies to the pursuit of pathogens—"the sentinels or sensors of the immune system," their discoverer, immunologist Ralph Steinman calls them. Once they've bagged their prey they mature into educators, mentors, motivators, counselors, and dispatchers. Now they will parent T cells, which rely on their instruction to make the transition from naïve recruit to trained soldier.

In principle, commanding a T cell to divide and take up arms should be no different from any other cellular directive. In practice, it's a daunting task. T cell receptors can't even see antigens until they have been cut into pieces and attached to carbohydrate-coated proteins encoded by the genes of the so-called major histocompatibility complex (MHC). Even then they're so short and stumpy and have such an anemic affinity for their antigen that the simple act of binding presents a formidable challenge. And once receptor and antigen *have* managed to join hands, that still isn't enough to activate the T cell's replication machinery or its munitions factories—the cell will only respond if in addition to the antigen signal, it receives a boost from a so-called costimulatory signal as well.

Dendritic cells take over where the thymus leaves off, completing the education of T cells by introducing them to their antigens and creating a climate conducive to intimate conversation. Expert taxidermists as well as hunters, within hours of capturing alien proteins they have butchered them, mounted the fragments on MHC proteins, and arranging the finished trophies for easy inspection by T cells—that's why their official job title is "professional antigen presenting cells." Ready to stop prowling for invaders and start looking for a partner, the antigen-decorated dendritic cells plant themselves in the lymph nodes, secrete chemokine "come hither" signals to draw in circulating T cells, and cook up a supply of the essential costimulator.

But their microbial victims are more than prey to the dendritic cells. They are also informants that supply dendritic cells with vital clues about the nature of the threat T cells will be facing. An elite corps of pathogen receptors—products of genes related to a fruit-fly gene that encodes, of all things, the receptor for the maternal signal that specifies the dorsal-ventral axis—interrogates the prisoner. From the combination of these Toll-like receptors (TLRs) engaged by microbial antigens, the dendritic cell can deduce what type of pathogen it's holding in its clutches—a Gram-negative bacterium, a flagellated bacterium, a multicellular parasite, a virus (in addition to ubiquitous bacterial antigens, certain TLRs also recognize double-stranded RNA, a hallmark of many viruses)—information it translates into directions for assembling an arsenal lethal to that type of pathogen."The beauty of the lymphocyte is its specificity. Lymphocytes have all these programs that can make them into killers or helpers, but how do they know what to do?" Steinman asks. "The dendritic cells tell them. They give the lymphocytes context, so they can choose the appropriate response."

But to realize these benefits, a naïve T cell must first locate a dendritic cell displaying the antigen that matches its receptors. Forsaking all others, it must then clasp its newfound master in an em-

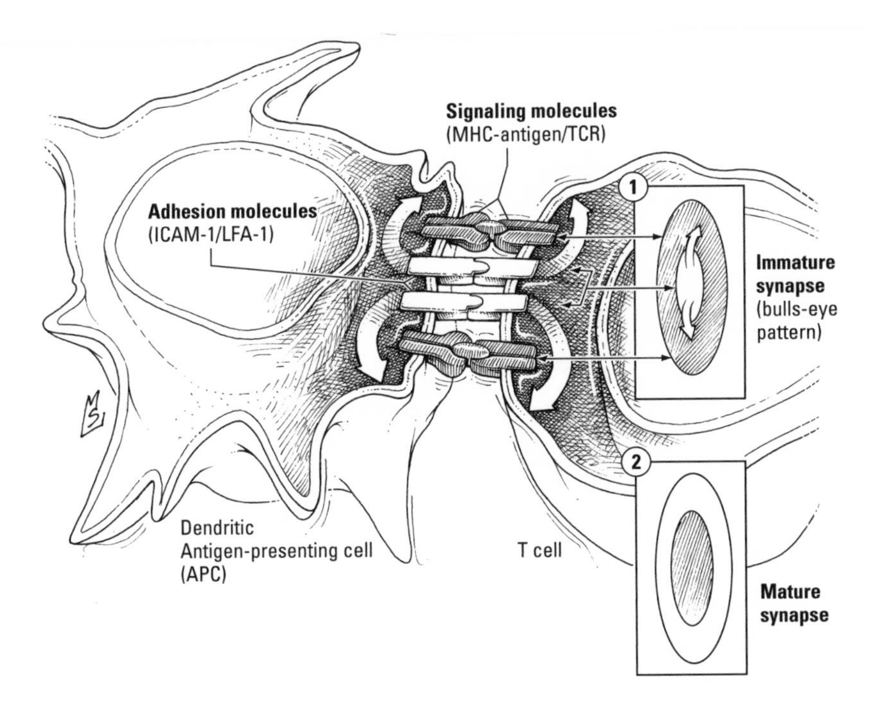

The immunological synapse. T cells are activated by intimate discourse with professional antigen presenting cells known as dendritic cells, an exchange regulated by an orderly arrangement of T cell receptors, antigen-MHC complexes, costimulatory signals and their receptors, and adhesion molecules known as the immunological synapse. When synapse formation begins, receptors and antigen complexes surround a core of adhesion molecules; as the synapse matures, these elements trade places—the receptor-antigen pairs move to the interior, and the adhesion molecules regroup around them.

brace intimate enough to bring antigen and receptor within working distance of each other, to hear the whispered costimulatory signals essential to activation. A precise connection for a private conversation? Sounds like a job for a synapse.

"In the late 1970s and early 1980s, researchers at the National Institutes of Health studying cytotoxic T cells (the ones that kill, rather than help) found that they used secretory mechanisms to kill infected cells. When the T cells interacted with a target, this apparatus was reorganized. It moved up against the target, polarizing the T cell, which then secretes [cytokines, for example] in a directed fashion," says Michael Dustin, an immunologist at New York University. Other researchers showed that helper T cells also used such directed secretion to talk to the B cells they activated. "Directed secretion reminded people of the neural synapse," Dustin says. As a consequence, the phrase "immunological synapse" began to creep into the literature to describe these intimate conversations between T cells and their partners, as well as speculation that the "immunological synapse" might be the basis of the T cell–dendritic cell interactions that initiated T cell activation.

By labeling proteins thought to be involved in immune synapse formation with fluorescent tags, scientists, including Dustin, demonstrated that the immunological synapse is assembled on the spot when the T cell contacted its signaling partner. Dustin calls it a "synapse on demand." Like synapses in the brain, it has a characteristic structure. Processed, immobilized antigen bound to MHC proteins, T cell receptors, costimulators and their receptors, and adhesion molecules are arranged in concentric circles, like a bull's-eye. The formation of this pattern is a dynamic process, a dance in which receptors and adhesion molecules exchange places.

Synapse making begins when T cell receptors detect and bind antigen presented on MHC proteins by the dendritic cell. At this point, receptor and antigen are in the outermost ring, surrounding a core of adhesion molecules. Then, perhaps in response to the interaction of the costimulator and its receptor, the receptors move to the middle of the synapse and the adhesion molecules shift to the outside, much as adhesion molecules flank the active zone in a neural synapse.

Initially, immunologists thought that the clustering of T cell receptors during synapse formation might be *the* critical factor initiating T cell signaling and activation. Then another difference between the immunological synapse and the neural synapse surfaced: T cell receptor signaling begins well before construction of the synapse is finished. So, "does T cell activation require this structure or does it just happen?" Dustin asks. He responds by saying that he believes the relationship between the T cell and the dendritic cell consists of three phases, one being the formation of the immune synapse. "During a brief initial transient phase, T cells keep migrating. They are actively receiving signals but continue to migrate," he explains. "About 6 to 8 hours later, they settle down with one dendritic cell and stay with that cell. During this sustained, or phase two period, which lasts for up to 24 hours, a T cell can still move, but when it does it stays with that one dendritic cell. This is the period of synaptic formation. After 24 hours or so, the T cell breaks out and starts migrating again."

Dustin argues that "cells gather useful information—although that information differs in quality and quantity—during all three phases. So there is something useful in stable interactions." That is, the immunological synapse is more than a platform for directed secretion—it is a stable environment in which participants can exchange critical information needed to regulate rather than initiate T cell signaling. For example, in 2003, he and collaborator Andrey Shaw demonstrated that receptor segregation within the synapse may be important for the fine-tuning of T cell activity to match the level of antigen. "The synapse downregulates T cell receptors if the stimulation is strong. If it's weak, then the receptor is concentrated in the center of the synapse and signaling is upregulated. It's like an iris. When you have lots of light, it closes, but when there's little light, it opens," he explains, adding that downregulation of receptors did not occur when synapse formation was disrupted. "It's an adaptive mechanism that keeps T cells from overexpanding or even generating a response strong enough to kill them. The mechanism protects them from overstimulation by downregulating the signal. And what happens at low antigen levels is that receptors are enhanced by the formation of clusters." Evidence also suggests that the synapse may facilitate the internalization of receptors in vesicles or promote and direct the exchange of cytokines at later stages in the T cell activation process.

"It's all an arms race between animals that want to live long and parasites and pathogens that want to live off them," Dustin concludes. Just as the neural synapse provides a mechanism for the precise, directed control of chemical signaling needed to coordinate behavioral responses—the basis of the adaptive capability of the nervous system—the extended and elaborate discussions between T cell and dendritic cell made possible by the immunological synapse underlie the response to diverse antigens, the capability at the heart of the immune system's incredible dexterity.

"The beauty of the system is [that] it knows how to adapt. It knows how to cope with pathogens," says Ralph Steinman. "That's its job."

Fancy new accoutrements like RAG proteins and mix-and-match receptors don't keep lymphocytes and cells of the innate immune system from chatting like social equals. And whenever they get together, the favorite topic of conversation is inflammation. During the initiation phase of the adaptive immune response, for example, signals released by mast cells and macrophages, particularly tumor necrosis factor (acting through the death-defying NF-κB pathway rather than the death-promoting FADD pathway; see Chapter 4, page 163), attract dendritic cells and warn them that the antigens they're consuming are likely to be pathogenic, while chemokines lure newly activated lymphocytes to the infection site. Conversely, helper T cells, specifically the subgroup designated "Th1,"exhort macrophages and neutrophils to keep on fighting. Once the combined forces of adaptive and innate immune systems have the invaders on the run, helper T cells of the Th2 variety secrete a different combination of cytokines to soothe macrophages, marking the beginning of the transition from attack mode to healing mode.

Conversations between the two arms of the immune system don't always have such a happy ending, however. The innate immune system may be set in its ways, but it rarely mistakes a normal constituent of its own tissues for a bacterial building block—its receptors, after all, evolved specifically to recognize features unique to microbes. Unfortunately, the adaptive immune system can't be this discriminating. Lymphocyte receptors recognize bits and pieces of almost any biomolecule from any species; to increase its reach, in other words, the adaptive immune system has had to sacrifice infallibility. While the innate immune system runs the risk of overlooking a deadly pathogen, it runs the risk of confusing a native protein with a legitimate threat. Should such a mishap occur, the normal crosstalk between the adaptive and innate immune systems can easily spiral into a deadly game of rumor, in which the two sides not only propagate but also amplify the initial misconception.

Among the millions of lymphocytes born in the bone marrow, some are certain to cross-react with body proteins. To prevent unwarranted attacks on body tissues, the adaptive immune system has established checkpoints to identify and purge these misfits from its ranks. Many, for example, are sent to their deaths during development, when their unnaturally high affinity for an illegal antigen gives them away. Others are caught and disarmed when they bind antigen in the absence of the costimulatory signal they need to start dividing and attacking or snagged and eliminated by dendritic cells that have picked up self antigens during their travels. Toward still others, the body adopts a live-and-let-live attitude, either because the antigen they recognize is so scarce they're unlikely to ever encounter it, or because it's sequestered in tissues with innate barriers that limit their interactions with the immune system. Should all else fail, disaster can still be averted through the intervention of innate regulatory mechanisms such as the calming words spoken by Th2 cells, which act as an emergency brake to contain the inflammatory response before it causes extensive damage. "Inflammation is a healing response, whether you like it or not. If you don't get a nice, normal inflammatory response, you don't heal," comments immunologist Cedric Raine. "Of course," he admits, "there's a potential for catastrophe. But there are also self-limiting events."

If you're one of the millions of Americans who suffer from rheumatoid arthritis, lupus, type 1 diabetes, Graves' disease (hyperthyroidism), or multiple sclerosis, you know that despite these safeguards, the human immune system, like the nervous system, can still be waylaid by misinformation and its own inherent adaptability. In these so-called autoimmune disorders self-reactive T cells have evaded mechanisms designed to kill or contain them, while failsafe mechanisms intended to quell the rebellion they instigate have broken down. As a result, defensive responses intended to protect you from pathogens have been trained instead on body tissues and the immune system has become locked in a relentless war with joints, nerves, internal organs, blood vessels, or heart valves.

Even tissues that take steps to minimize their contact with the immune system are not safe from self-destructive attacks launched by confused T cells. Take the brain, for example.

Axons, like wires, conduct electricity more effectively if they're surrounded with a layer of insulation. To that end, many nerves are enveloped by supporting cells known as Schwann cells (in the peripheral nervous system) or oligodendrocytes (in the central nervous system), which wrap themselves around the nerve to create a layer of insulating material called myelin. Myelin contains both lipids and proteins, the best known of which are myelin proteolipid protein, myelin-oligodendrocyte glycoprotein, and a complex of seven closely related proteins known collectively as myelin basic protein.

So-called autoreactive T cells that recognize myelin proteins, particularly myelin basic protein, can be found in the blood of even normal healthy people, part of the contingent of malcontents that body has chosen to ignore. As long as they remain innocent, they cause no problems. And innocent they are likely to stay, for the myelin of the brain is shielded behind the highly selective barrier formed by the brain vasculature, which limits the transit of naive lymphocytes, along with all but the smallest, most lipid-loving molecules.

Occasionally, however, a perverse twist of fate—infection with a virus containing a protein that's a dead ringer for myelin basic protein is one popular candidate—activates myelin-reactive T cells. Quiescent, autoreactive T cells were excluded from the central nervous system. But these active "autoaggressive" T cells are at liberty to ignore all barriers. "Step aside! Might be an infection!" they roar as push their way, armed and spoiling for a fight, into the brain or spinal cord. Instead of microbes, however, they find myelin. Spewing cytokines and chemokines that bring ravenous macrophages and more autoreactive-turned-autoaggressive T cells running, they precipitate an inflammatory immune reaction that annihilates the cells responsible for myelin production in the brain, spinal cord, and optic nerves. Without myelin, nerve conduction slows to a crawl. Weakness, numbness, or loss of vision in one eye are often the first symptoms of the destruction, followed by a pattern of alternating remission and relapse and progressive disability. Dubbed *sclérose en plaques disseminées*—multiple sclerosis—by Jean Martin Charcot, the French neurologist who first described this autoimmune catastrophe in 1868, MS, as it's known colloquially, affects nearly 1million people worldwide, the majority women, two-thirds between the ages of 20 and 40.

Antigen recognition in multiple sclerosis skews the helper T cell ratio in the Th1 direction. Instead of defusing the conflict by dousing the participants with anti-inflammatory cytokines, like any socially responsible Th2 cell would do, Th1 cells shout incendiary words certain to feed the fire and increase the damage to innocent cells: the inflammatory cytokines interleukin (IL)-2; interferon-gamma; and the proinflammation, proproliferation, prodeath signal TNF-α. The release of these signals heralds the beginning of an acute inflammatory episode; in fact, it may actually precede the attack by several weeks. TNF-α is an especially baleful word, a synonym for "disability": the higher the level of TNF-α in the cerebrospinal fluid, the more impaired an MS patient is likely to be. What's more, it's a bad sign for the future—one study has found that the higher TNF levels over a two-year period, the more rapid the rate of disease progression.

In addition to its other meanings—"Drop dead," for example, or "come here"—TNF-α says "knock, knock" to the endothelial cells lining brain blood vessels. In response, endothelial cells don adhesive molecules (just as they do to capture macrophages) that cue bloodborne aggressive T cells to slow down and roll, looking for a door to the inside of the brain. A gaggle of chemokines then show them where to find the friendliest integrins, and they stop to chat. Talk leads to more intimate relations with endothelial cells, so smitten they go slack-jawed, allowing T cells to squeeze through the blood vessel. "Charge!" shout the chemokines, and the T cells fall on the hapless oligodendrocytes. Half-chewed myelin generated by the attack provides a fresh infusion of myelin antigens, while the relentless battle cries of T cells draw even more recruits to the site of damage, perpetuating or exacerbating the autoimmune response.

Once scientists learned the secret words traded by cytokine-inspired T cells and endothelial cells willing to pull them out of the bloodstream, they could interrupt the conversation with a word of their own—"natalizumab." An antibody to the integrin receptor VFA-4, found on the surface of active myelin-reactive T cells, "natalizumab" sounds so much like "V-CAM1," the endothelial cell adhesion molecule that matches VFA-4, it's no wonder T cells confuse the two. By the time they realize "natalizumab" is actually a nonsense word, not a handhold on the blood vessel wall, they've been swept downstream, far from the myelin-bearing cells they were intent on attacking.

Neurologists had high hopes for natalizumab, after a 2-year clinical trial in more than 900 MS patients showed that its ability to block the passage of myelin-reactive T cells from blood to brain reduced the relapse rate by 66 percent compared to patients receiving a placebo. But luring T cells away from their jobs also undermines the alliance between adaptive and innate immune mechanisms that has proven so effective in the fight against crafty pathogens, tamping down autoimmune reactions at the risk of increasing susceptibility to infection. Just three months after natalizumab—Tysabri, as it was known in the marketplace—was approved by the Food and Drug Administration under a new policy designed to expedite the introduction of drugs to treat potentially life-threatening conditions like MS, sales were suspended in response to reports that two patients had developed a rare viral infection typically seen only in severely immunocompromised patients. "I'm shocked that it happened so soon," said one expert, Lawrence Steinman, in an interview with *The New York Times*. "But I knew it was going to happen sooner or later."

Paul Ehrlich, whose ideas about the interactions between antibodies and hypothetical "side chains" on the surfaces of pathogens helped frame the concept of receptors, called the ravages of a confused immune system *horror autotoxicus*. Expanding the reach of the immune system, to ensure it can deal with the pathogens of tomorrow as well as those long familiar to the human race, carries with it the implicit risk of debilitating or even, as Ehrlich noted, horrific results. Like memory, immunity can overreact to the wrong information. Yet "survival is impossible without vigilant defense against attack and injury," and that vigilance can be had only at the price of accepting the peril it entails.

"'Change is suffering' was the insight that founded Buddhism," writes Stewart Brand. "We hate change. Ever since the big easy chair was reupholstered it's not as comfortable anymore. And we love change. Let's just *redo* the kitchen! To change is to lose identity; yet to change is to be alive."

THE VIRTUAL CELL

The receiving line includes distinguished professors and humble graduate students, the well known and the unknown, scientists who spend their days peering at computer screens and others who spend theirs peering through microscopes, representatives of government agencies, a sampling of corporate sponsors, and a contingent from a respected scientific publisher. It's the 2002 opening reception for the second annual meeting of the Alliance for Cellular Signaling, a consortium of investigators with a common interest in the language of life, and the man everyone is waiting to greet is Alfred G. Gilman, the Alliance's creator, director, taskmaster, and muse.

Regal in his formal dark suit and crisp white shirt, Gilman looks the part of the elder statesman as he greets his guests. Launched in September of 2000—two decades after he added "G_s" to the lexicon of signaling molecules—the "AfCS" is Gilman's idea for making sense of the multivolume tome the dictionary has become. Over a 10-year period, the 80-odd scientists who currently make up the AfCS intend to compile a comprehensive "parts list" for every signaling pathway in two types of mammalian cells: the mouse B lymphocyte and the myocyte, or muscle cell, of the mouse heart. But that's only the beginning. The AfCS also hopes to understand how these proteins interact to process information, "the big question of how they all work together as a system." Then, armed with this information, they plan to design a computer model of each cell that will re-create the network of interconnected signaling pathways and use this "virtual cell" to explore the dynamics of the signaling circuitry as well as aid in the design of new drugs to correct defects in that circuitry. Oh, and along the way, they hope to change fundamentally the way science is done in the twenty-first century.

"Sequencing the genome is enabling us to think about bigger questions in unbiased, nonhypothetical ways," Gilman noted in an interview at the time of the launch. But solutions to bigger questions, he believes, are more likely to come from the collaborative efforts of many scientists, pooling resources, sharing data, and developing the analytic tools to manage and mine those data than from the time-honored style of research based on individual scientists working by themselves, for themselves—solutions, Gilman argues, that can be facilitated by modern communications technologies like the Internet.

To build a virtual cell, therefore, he has built a virtual laboratory. Eight research teams, located in San Francisco, San Diego, Pasadena, and Dallas, carry out the actual experimental work and develop the software needed to store and analyze the data they generate. An additional 41 participating investigators serve on committees that oversee the work of the Alliance laboratories. Everyone communicates regularly via teleconferencing and the Internet, enabling them to coordinate their efforts as if they worked in the same building. Finally, approximately 1,500 subscribing members will contribute by authoring so-called Molecule Pages, peer-reviewed descriptions of information about signaling molecules the AfCS will catalog and publish as an electronic database freely available to all members of the scientific community.

Big science takes big money. The annual budget for the project totals around $10 million, much of it coming from a $25 million "glue grant" awarded to the AfCS by the National Institute of General Medical Sciences, as part of an initiative to support and encourage such collaborative endeavors. The balance will be made up by funds from the National Institute of Allergy and Infectious Diseases, the National Cancer Institute, paid subscriptions from a consortium of pharmaceutical companies, and several private donors. In addition, a group of biotechnology companies contribute to AfCS research by supplying equipment and reagents essential to specialized, state-of-the-art techniques.

"This is not an easy operation to run. You could think of it as a small company spread all over the place," Gilman concludes. "Participants do not have proper, conventional academic motivation. We're not providing money for their laboratories or publication opportunities. People are part of this because it's an interesting—and important—project."

The 2001 meeting was all about organization—the design of experiments, the standardization of procedures for recording, transmitting, and analyzing data. At this meeting, AfCS investigators will unveil the first actual results of their efforts. The mood is expectant, excited. Members greet one another as effusively as if they were attending a high school reunion rather than a scientific meeting. It's been a hard year, everyone agrees. And the biggest problem hasn't been the experiments or the data analysis, but the struggle to get people more accustomed to competing with one another to collaborate instead.

"It's been a process of organization, as well as science, sorting out the hierarchy," says one member. Another, one of the computer experts responsible for organizing the Alliance database, agrees, noting that despite their reputation for fussiness, "bench" scientists are anything but consistent when it comes to record keeping. "One of the hardest parts has been teaching scientists to think about their data before they do an experiment, rather than afterward," he explains. "They're not used to standardized procedures. You have a notebook and one guy wants to keep it his way and another wants to use his own way. One wants to write in red ink and one in blue ink"—or the electronic equivalent anyway. A third complains that some tasks still haven't been converted to an electronic format and must be carried out manually. "It's like a big ocean liner, out in the middle of the ocean. At one end there's an engine powering the boat along, and at the other there's still some guys rowing," he says as he spears another shrimp roll from a passing tray. "And I'm just glad it floats."

Gilman sums it up: "We are both experimentalists and experimental subjects. This is an experiment in how to do science as well as an opportunity to find out about signaling systems. There are people watching to see if we can make this model work. The world, in general, is a little bit skeptical of big science. And the world is watching. We want to deliver what we promise."

THINKING OUT OF THE "STRAIGHT LINE" BOX

In 1992 the journal *Trends in Biochemical Sciences* published an update to the "black box" insulin cartoon that diabetes expert Ron Kahn found so compelling. The new version also depicted a cell and its insulin receptor. But now instead of black shading, the area inside the cell contained a bewildering array of insulin substrate proteins, second messengers, kinases, adaptors, glucose transporters, and transcription factors, connected by a labyrinth of intersecting lines and arrows. At the bottom the new caption reads, "Now we *really* understand insulin signaling."

Do we?

Over the past four decades, biologists have compiled an impressive list of signaling molecules, from the chemotaxis proteins of *E. coli* to the plethora of insulin-related proteins and the dozens of neurotransmitters ferrying messages in the brain. They have demonstrated how the loops and notches of protein domains can link protein to protein or pathway to pathway. They have cataloged patterns and outlined the rules of cellular grammar. They have demonstrated how signaling mechanisms can connect environment and genome, strike a balance between excess and shortfall, reconstruct as well as regulate.

The achievements of signaling research are representative of our success at analysis—tearing things apart—and our urgent need for synthesis, as scientists have also come to recognize that cell-cell communication is more than linear information transfer. In the living organism, "signaling pathways interact with one another and the final biological response is shaped by the interaction between pathways." Like the fiber optic systems that route telephone calls or the hardware that shuttles data inside your computer, biological communication is the product of integrated circuits operating as a network, and the collective behavior of that network, not the sum total of the activity in myriad individual units, is what determines the cell's responses to external signals.

"The term 'pathway' implies a linear structure—you flip a switch and a particular biological process happens at the other end. And that's clearly flawed," says cancer researcher Gerard Evan. "Signaling pathways can never be linear because they would never work," he continues. "There aren't enough genes to encode all the combinations you would need. Any time you couple two pathways and it confers a selective advantage, that connection will be retained. So what you get after billions of years of evolution are connections.

"We're primates, so we like to put things in boxes. But evolution is like going into your basement and sorting through the rubbish to see if anything is useful."

Evan cites the transcription factor Myc as an example of a protein that has been favored by evolution because it could be recycled, mixed and matched with other proteins to form multiple signaling pathways. "Myc is an integrator, bringing together all the things you need to proliferate. Any gene coupled to *myc* that enabled an organism to build tissues more effectively would be selected for. So you get a bewildering proliferation of pathways.

"Process comes from the way pathways interact, not specific proteins," Evan concludes. But it's one thing to acknowledge the importance of interactions and networks, and another thing entirely to investigate signaling from a network perspective.

"Complexities can only be understood by constructing quantitative, often mathematical models," writes former *Nature* editor John Maddox in his book *What Remains to Be Discovered*. According to systems biologist Eric Werner, scientists realize three benefits when they construct such models to explain complex systems. First, models allow researchers to evaluate their progress, because the development of models exposes gaps in their understanding of biological systems. Second, models can be used to inform future experimentation: "Models . . . can be used to plan experimental strategies," while "cycles of modeling and experimental validation gradually result in the convergence of the model's predictions with the measured parameters of the natural biological system." Finally, Werner argues, the quantitative rigor demanded by model building exposes fuzzy thinking; as he puts it, models "force a new perspective on the subject matter. One can no longer tolerate intuitive, vague models. One is forced to look at the consequences of theoretical assumptions."

Unfortunately, as Maddox notes, biologists have been slower than other scientists to appreciate the value of model building. Their reluctance stems in part from the fact that many are unfamiliar with the specialized mathematical and computer skills that modeling demands or are disconcerted by the idea of experiments conducted "in silico," on computers instead of at the lab bench. But the cult of the individual researcher is also a barrier. An integrated biology that seeks to combines database management, data mining, and network analysis with experimentation, especially experimentation that involves state-of-the-art molecular biology and imaging techniques, is "big science," made possible only by pooling time, effort, and expertise. If biologists are to construct models as well as perform experiments, it will require not only a paradigm shift, "refocusing on systems, rather than components," but also a cultural revolution—an overhaul of the field that Gilman and his collaborators hope to pioneer.

Spoken language is more than strings of sounds or even combinations of words. As architect Christopher Alexander says, "Each word carries the whisper of the meanings of the words it is connected to"; that is, the meaning of the entire sentence depends not only on the words themselves but also on the relationship between words. In the living language of life as well, the relationships between chemical words add a subtlety that transcends the individual elements. We can parse pathways into proteins, but to understand the coordinated behavior of real organisms, we must also understand how meaning emerges from the confluence of proteins and the interactions between pathways.

If you're wondering what all this number crunching and paradigm shifting has to do with you, there's a practical benefit to what the AfCS is attempting to do as well—it will be good for your health. Chemotherapy, Paul Ehrlich's idea that cures could be found in chemical agents with a specific affinity for the "side chains," has been one of the most productive concepts in medicine, and the interactions of experimental compounds and signaling proteins, particularly receptors, represent the cornerstone of modern drug discovery. Today, the overwhelming majority of drugs in the modern pharmacopoeia act by talking to cells in their own language. The discovery of new signals and new receptors, as well as the intracellular relays they engage, has offered new targets for pharmaceutical researchers, opportunities that are already beginning to yield more effective and less toxic medications—and, at the same time, have revealed the pitfalls of concentrating on individual receptors or kinases instead of the larger picture, particularly the interrelationships between signaling pathways.

The general public may view cancer as "a modern-day plague" and oncologists may see it as "a legion of recondite diseases whose diversity complicates therapy," but to a cancer patient, it is "a terrifying alien entity invading his body and treatable only with medicines of medieval harshness and dubious efficacy," writes Gerard Evan. Indeed, for decades, cancer treatment has focused on eradicating tumor cells by brute force, with surgery, radiation, or powerful drugs; "slash, burn, and poison," breast cancer authority Susan Love calls it. The side effects of bludgeoning cancer into submission can leave patients wondering if the cure isn't as bad as the disease.

But in 1998, Genentech introduced a different kind of cancer drug, an agent that attacked cancer by silencing its propaganda machine rather than by mounting a brute force assault. The drug, Herceptin, is a monoclonal antibody to a domain in the protein encoded by the *her2/neu* gene, related to the receptor for epidermal growth factor. Malignant cells in some breast cancer patients contain extra copies of the *her2/neu* gene; these cells proliferate extravagantly because they're saturated with extra receptors as a consequence (as many as 1.5 million receptors per cell, compared to about 50,000 on normal cells), leading them to imagine they are being bombarded with growth factor. By binding and blocking these receptors, Herceptin turns down the volume—without harming normal cells.

In 2001, Novartis followed this success with Gleevec, another drug that works by talking cancer into remission. In patients with chronic myelogenous leukemia (CML), Gleevec silences an aberrant kinase, Bcr-Abl, created when segments of chromosomes 9 and 22 accidentally switch places. In patients with a rare type of intestinal cancer, gastrointestinal stromal tumor, Gleevec blocks a mutant receptor tyrosine kinase called c-kit. Finally, it also inhibits the platelet-derived growth factor receptor tyrosine kinase; as a result, clinical trials are evaluating its effectiveness in cancers with mutations in this receptor. Gleevec is breathtakingly effective—90 percent of patients with CML who take the drug in the early, chronic stage of the disease have their cancers go into remission, as do 60 percent of patients with gastrointestinal stromal tumor. In addition, because it targets defects specific to cancer cells, Gleevec, like Herceptin, spares normal cells, minimizing the incidence of side effects.

A third signaling oriented drug, AstraZeneca's Iressa, was approved in 2003 for the treatment of non-small-cell lung cancer. Like Herceptin, Iressa shuts off the relentless screaming of a corrupted epidermal growth factor signaling pathway, in this case due to mutations in the growth factor receptor itself. Rather than smothering superfluous receptors, however, this agent works inside the cell to prevent the EGF receptor from phosphorylating itself. Without the addition of phosphate, the receptor can't communicate with its intracellular signaling relay, and growth factor–induced responses grind to a halt.

"The war on cancer was started by Nixon, but we're only making progress now," says cancer researcher Philip Beachy. "As we understand more about the cells involved in tumors we can target those cells instead of using blunt instruments that kill all dividing cells. In the future, oncology will have a different set—a more specific set—of therapeutic tools. Instead of cytotoxic agents, we'll have mechanism-based agents."

Continued progress in developing a "more specific set of therapeutic tools" depends critically on a better understanding of the interlocking signaling mechanisms governing the progression to malignancy. A challenge, yes, but not an insurmountable one, argues another cancer expert, Gerard Evan. "We have only two options to collapse the cancer platform," Evan writes. "One is to attack the lesions that drive tumor cell proliferation"—such as growth factor receptor mutations. "Alternatively," he continues, "we could reinstate the defective apoptosis, whereupon the tumor cell should die from the apoptotic depredations inflicted on it by its driving oncogenic lesions." He acknowledges that this second option is currently hampered by our relative ignorance of the genetic accidents that derail the suicide program in cancer cells. Still, he points out, we may have already begun to correct the failure to die in spite of ourselves, noting that drugs designed to interfere with growth factor signaling not only block the compulsion to proliferate but also eliminate essential survival signals. With a better understanding of suicide signaling—and dozens of antiapoptotic drug candidates under investigation—the future may lead to more deliberate approaches to rebalance cell proliferation and cell death.

Signaling drugs like Herceptin, Gleevec, and Iressa are not the latest miracle cures. Only 25 to 30 percent of breast cancer patients have the defect leading to runaway expression of the HER2 protein that is the target of Herceptin (and not all of these patients respond to the drug, possibly because HER2 signaling is more important in some cancers than in others). An even smaller number of lung cancer patients—about 10 percent—have the precise growth factor mutation targeted by Iressa. Gleevec is highly effective in the early chronic phase of CML, but patients with advanced disease in the acute, or "blast-crisis" phase often fail to respond, and some who are respond initially relapse when their cancer cells spawn another mutation that allows them to ignore the drug. But their unique mechanism of action makes these agents the first real advance in cancer chemotherapy in decades of research. Medications that finally hurt cancer more than they hurt the patient, they are leading oncologists to suggest—cautiously—that drugs like Gleevec might turn cancer into a chronic but manageable disease.

At the same time pharmaceutical researchers are making progress in the war on cancer with new drugs that target signaling proteins, they are running into roadblocks in more traditional areas of signaling research. Take drugs that act by stimulating or blocking signaling at G protein–coupled receptors. This category includes many of the industry's most successful drugs, from antihypertensives to antidepressants. Yet progress in identifying new drugs that target the signaling pathways headed by these receptors has slowed to a crawl, in part, some experts believe, because drug discovery research has focused so intensively on identifying compounds that bind more and more specifically to particular receptors, without considering the networks of intracellular signaling proteins that take over once messages cross the plasma membrane. "When success does occur, many come out with very similar molecules targeting the same receptors. Metaphorically speaking, it is as if everyone is looking under the same lamppost to find the key to the biological problems being considered, because they only feel comfortable in that well-lit environment," says pharmacologist Michel Bouvier, in a recent roundtable discussion published in the journal *Nature Reviews Drug Discovery.* "We are still at the stage of identifying the various partners that can be engaged by the receptor, and we are only just getting the first glimpses into their potential roles in controlling signaling selectivity and efficacy," he adds. "Undoubtedly, this will dramatically change the way we approach signaling selectivity, and will hopefully provide new insights into how we can modulate GPCR signaling in a selective manner."

That's why pharmaceutical companies are joining government agencies and the private sector in bankrolling the AfCS—they recognize that the more we know about the complexities of signaling, the more successful they will be at identifying and exploiting new therapeutic opportunities.

BUILDING AIRPLANES IN THE SKY

The mood is still ebullient when Gilman takes the stage the next morning. "You came back!" he announces and everyone laughs. He leads off with a summary of the AfCS strategy and goals: "How does a signaling mechanism work? What are the components of our system? Who interacts with whom? How does information flow through the system? And finally, can we end up modeling this system and predict the behavior?" Then, to demonstrate what a tall order this actually is, he shows a video clip, a commercial for the computer firm Electronic Data Systems.

"Some people like to climb mountains," announces a man who appears to be a sky diver, suspended in midair from his parachute. "I like to build airplanes . . . in the air!" The clip goes on to depict the thrill of victory he and his colleagues experience from bolting an airliner together at 40,000 feet, interspersed with glowing testimonials from enthusiastic passengers. "When I see that look on a little kid's face, that's all the faith I need," the narrator concludes. "Whooohooo!"

"We are building an airplane in the air," Gilman says as the lights come up. "As you see, it can be done. And the look on that little kid's face *is* all that matters."

Getting back to business, he assures the audience that "progress has been substantial, especially in the last couple of months." They'll be hearing details of the methods Alliance laboratories have developed for preparing and propagating each of the two cell types, he notes. And he explains that much of the data they'll see will come from so-called ligand screens intended to flesh out the "parts list" for the signaling pathways of each cell and "measure the spectrum and pattern of responses." Basically, the major goal of this part of the work, Gilman observes, "is to collect empirical information for use at later stages." For example, investigators want to know if all ligands that work through G protein–coupled receptors behave similarly or if they differ and, if so, how. In addition, the ligand screen will attempt to identify modules, groups of signaling molecules that work together as teams in multiple signaling pathways (e.g., MAP kinases). But no peeking, he adds. "I'm not going to take the fun away from the people who have done the work."

One by one, representatives of the Alliance labs take the stage and review the year's work, beginning with the laboratories in charge of the B lymphocyte, which, according to B cell team leader Henry Bourne, of the University of California, San Francisco, is "alive, well, and rarin' to go!" Their methods of isolating and tending these cells have been perfected to the point that they can routinely prepare nearly 20 billion per week, he notes. Thanks to this productivity, they've screened 43 ligands so far, charting alterations in gene expression, the phosphorylation of proteins. And although AfCS members are committed to working with primary cells (that is, cells removed and cultured from animals, as opposed to established cell lines), just to be on the safe side, B lymphocyte investigators have hedged their bets by setting up and screening a B cell line, WEHI-231. "If primary cells can't be tractable, the hell with them," Bourne threatens. "We should play to our strengths." Ideally, he adds, the primary cells will be the "gold standard"—the definition of how signaling transpires in actual living cells—while the WEHI-231 cells will be the "workhorse," for large-scale studies.

If anything, the B cell screens have been too successful, Bourne says, generating a "tsunami of data." As a consequence, this team's greatest challenge has been archiving and analyzing all these data, a task, he admits, that "has been like giving birth to barbed wire."

The scientists working on the cardiac myocyte should be so lucky. Chosen as much for their esthetic appeal as its biological importance—"It's a beautiful cell to do business with, a beautiful cell in culture, a beating cell," Gilman notes—myocytes are as fussy as tropical fish when they're placed in a culture dish, and they live about as long. The dilemma this team must answer is: Can the myocyte survive and, more importantly, can they survive the myocyte? Still, Myocyte Committee Chair Jim Stull is cautiously optimistic, reporting that his team can maintain the cells for at least 24 hours without death or degeneration. With a preparation that finally lives long enough to be tested, he says, they have a "green light to begin the single ligand screen for this cell type."

At the end of the day, Henry Bourne sounds a note of caution.

Success, he argues, raises the specter of "'magical inductionism,' the fallacy that if you simply collect enough data, knowledge will ensue." To "take the magic out of it," he recommends that the group concentrate on one part of the map, rather than trying to take in the whole continent—he recommends the pathways that use PIP3 as a second messenger. And by the way, he notes, some more powerful analytic tools would be nice, too.

"You wanna see the Grassy Knoll? You come this way—I'll take you on a tour," offers a seedy-looking character in a dirty green T-shirt. Dinner that evening is at the infamous Texas Book Depository—on the seventh floor, not the sixth, where Lee Harvey Oswald took aim. Here, scientists collect in groups to discuss the day's presentations, people they have known only by teleconference until today. "It's like meeting a celebrity," enthuses one. "You've only seen them on TV and then here they are in person." To which his colleague responds, "Yeah, people looked like I thought, except they were different sizes." Another trio is discussing the best watering holes in the Far East. Eight or nine are hogging the cheese and crackers during a lively discussion of the solubility of membrane proteins.

Asked for their reactions to the Alliance's progress, they agree: it is like building airplanes in the sky, and, yes, imagining the looks on outsider's faces when they've built their virtual cell will be worth it. One confides that most of the results were generated in the past few weeks. "It really was like giving birth to barbed wire," he says. For members Temple Smith of Boston University and Paul Simpson of the University of California, San Francisco, however, the most impressive thing isn't the data, but how the Alliance itself has succeeded in working as a network. "Look around the room," says Smith. "A diverse group of scientists can overcome barriers and communicate and work together: physicists, chemists, bioinformatics specialists. Our biggest challenge now is that there are a zillion attractive things to do but not a zillion dollars, brains, or people."

By 2003, the biggest challenge would be keeping the project from running aground.

The crisis was strictly technical, not organizational. Both of the cells chosen by the steering committee at the beginning of the project had failed to live up to expectations. "To be considered, a cell had to meet several criteria," Gilman explains. "The most important was that we wanted to work with real cells, not cell lines. They had to be mammalian cells, from an experimental species with a sequenced genome—meaning, at the time, that it had to come from the mouse. It had to be large enough for microscopy and manipulated readily. It had to be normal—no cancerous cells, no matter how readily they took to cell culture. And it had to 'do something.' But we left off one criterion that should have been on the list. And that was easy."

An observer could have foreseen the demise of the finicky, undependable cardiac myocyte; always difficult, "we were never quite certain about it," Gilman admits. "The preparation was tortuous. We were making progress, but we weren't sure it was worth all the effort." The real disappointment, however, was the B cell. Able-enough performers, in the end they presented one insurmountable difficulty.

Gilman explains that everyone had agreed that the technique known as RNA interference, a way of "knocking out" proteins (singly or in combination), was a "must-have" technology. "One of our goals was to use RNA interference to disrupt pathways, even though, at the time, the technique worked only in worms and fruit flies. We were going to light candles for the people trying to do this in mammals. And they succeeded. RNA interference does work in mammalian cells—just not ours." Despite heroic efforts, investigators could not get the technique to work in the B lymphocyte. Worse, they could not get the technique to work in their back-up option, the WEHI-231 cell—a disappointment Gilman calls the "ultimate ironic shaft."

Reluctantly, the AfCS Steering Committee decided to give up

on primary cells, change course, and choose a cell line to be their new "gold standard." With the 2003 annual meeting just two months away, the AfCS laboratories went to work, collecting enough data for participants to evaluate the merits of three candidate mouse cell lines: two derived from a white blood cell of the phagocyte clan, the macrophage (J774A.1 and RAW264.7) and one from the pituitary gland (AtT-20). The winner, hands down, was RAW264.7. Like most cell lines, it was easy to grow and stable in culture. It submitted willingly to RNA interference. It came from a mouse, so the tools and reagents the Alliance had already developed could be recycled. Finally, being a macrophage, "it was a performer—it secretes cytokines, it engages in phagocytosis, it moves."

"Some might fear we wasted a lot of time," Gilman observes. "But in fact, we learned how to do a lot of things. In particular, we had built the infrastructure."

Actually, they built an airplane. To date, double ligand screens, involving 24 signals and four parameters—calcium, cyclic AMP, 21 phosphorylated proteins, and 18 cytokines (facilitated by the development of an assay measuring all 18 simultaneously)—have been completed. Following Henry Bourne's suggestion, Alliance investigators limited more detailed studies of RAW264.7 signaling pathways to a single "X module" encompassing the responses of calcium and PIP3 to three ligands. In the first stage of this project, they compiled a parts list of about 200 signaling proteins. Next, using the coveted RNA interference technology, they began to examine connections between pathways. Finally, by comparing the data they have generated with "legacy" data from the literature, using state-of-the-art software, they are in the process of verifying "one of the most detailed maps of pathways leading from the activation of receptors . . . to stimulation of calcium"—our first glimpse of the signaling circuitry in a living cell.

A century ago signaling investigators translated the first words of the language spoken by cells. Forty years ago, they learned how to track receptors with radioligands, and began to speak in two-word phrases. By the beginning of the new millennium they had pieced together sentences, even reconstructed portions of paragraphs. And now, four years later, they have a small and unfinished yet exquisite book to show for their efforts.

Labor Day, 2004. Gilman is working on the renewal of the glue grant that forms the major portion of the Alliance's funding. It's a time to review the big picture, to look back on what has been accomplished and to map out what remains to be done.

"From an experimental point of view, we're just now getting to the fun stuff. But this social experiment is working very well. We've proven that we can work together and that modern communication technologies make long-range projects viable. People do have to have an appropriately selfless attitude. As far as that goes, there's some correlation with age and, as a result, the degree of security they feel. It's harder for younger people. But what they gain is a crack at solving a really big problem."

"Five years isn't a very long time," he admits. "We can't claim any huge victories yet. What we do see are the substantial interactions between pathways that everyone suspected were there but had never been documented. The level of complexity is huge."

As for the next five years, Gilman is resolutely optimistic. "What can I say?" he concludes. "We'll make a lot of progress. We have a gold mine of data, although it's not obvious yet where all of the gold is. And in the end, we've generated a hypothesis machine, one that can and should inspire lots of questions."

A SYSTEM OF CENTERS, THE PRINCIPLES OF POEMS

Ever since the publication of *A Pattern Language* more than 30 years ago, Christopher Alexander has felt that his theory was not quite complete, that he was missing something vitally important. Now, he says, he knows what that something was—the big picture.

In his latest work, *The Phenomenon of Order*, the first of four books intended to constitute an "essay on the art of building and the nature of the universe," Alexander argues that beautiful buildings—and living things—display "wholeness," a quality he attributes to entities he calls "centers." He defines a center as "a physical set that occupies a certain volume in space and has a special marked coherence" and emphasizes that centers not only create the wholeness but are created by it; they seem to emerge from it as intersecting elements of a larger pattern, rather than distinct units. In contrast to the way we are used to defining space—the proverbial four walls—centers have fuzzy boundaries, are more fluid; they do not necessarily correspond to items we can name. But this imprecision is unimportant because the details of each center do not matter as much as their interrelationships, it is this "system of centers" in a building, a scene, or a community that creates wholeness.

As an example, Alexander asks readers to consider a country house, surrounded by a garden:

> I notice the sunny part of the garden itself as a space. The place where the roses are climbing near the kitchen catches my eye. The path to the front door, and the steps from the back porch, and the door itself, the door of the house, all work as a unit. . . . The sunshine and the roof edge, with the rafters repeating under the eave together form a pattern of light and shadow which leads my eye, and forms a boundary of the house against the sky.
>
> All this is much more like a pulsating unity than the "conceptual" or intellectual image of the house. In our conceptual picture of the house, we have things called street, garden, roof, front door, and so on. But the centers or entities that hit my eye when I take it all in as a whole are slightly different. I see the sunny part of the garden where the sun is falling on the lawn as a center—not the entire "garden."

After careful examination, Alexander has identified 15 fundamental structural properties he believes are common to things exhibiting wholeness, things with "strong centers" that "have life." Number eight is something he calls "deep interlock and ambiguity," in which a center and its environment, or that center and another, interdigitate, forming the sort of pattern you might see in a house with an expansive wraparound porch, fine lace, or a Persian carpet. Number 15 is "not separateness." A building with this quality melds harmoniously with the world around it, a space, such as a garden, with it is filled with things that connect to one another in some way; each part "melts into its neighbors."

"Nature too is understandable in terms of wholeness . . . centers, wholes, and boundaries occur repeatedly throughout the natural world," Alexander writes. As a result, "the fifteen properties appear as geometric features of the way that space is organized in nature." Deep interlock and ambiguity, for example, appear in the pattern of a giraffe's coat or the involutions of the brain. And not-separateness is an integral feature of any ecosystem, from a backyard garden to a rain forest, and "corresponds to the fact that there is no perfect isolation of any system"—even a system as small as the collective signaling pathways in a single cell.

An architect with more technical virtuosity than insight can "make buildings by stringing together patterns, in a rather loose way," writes Alexander. A house cobbled together this way may be a shelter, but it is not a home; its walls and windows and doors are no more than an "assembly of patterns"; it stands out painfully from its surroundings. On the other hand, when patterns are integrated—for example, by locating "family space" within the kitchen, replacing a solid wall with interior windows, adding a terrace, or planting a garden—deep interlock makes the finished building "very dense; it has many meanings captured in a small space; and through this density, it becomes profound." In contrast to the aridity of cookie-cutter suburban houses and offices composed of cubicles, the visual and emotional impact created by superimposing design elements and erasing the boundary between the house and its surroundings adds a complexity that can transform ordinary living spaces into "buildings which are poems."

Living things, too, have deep interlock and ambiguity, display not-separateness. The lines of communication within our cells, for example, are not just strings of proteins but inseparable interlocking patterns compressed into a space far smaller than the tiniest speck we can see with the naked eye. The exchange of chemical messengers within these networks weaves a tapestry of larger patterns that link cell to cell, tissue to tissue. From these intricate and meaningful conversations between nerve and muscle, bone and blood, mesoderm and ectoderm, cell cycle and death machinery emerge bodies that are dense with meaning, organisms that are also poems.

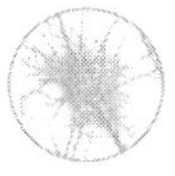

NOTES

INTRODUCTION

p. 2 **a veritable city of an organism:** J.T. Bonner, *First Signals: The Evolution of Multicellular Development* (Princeton, NJ: Princeton University Press, 2000), p. 56; S.F. Gilbert, *Developmental Biology*, 7th ed. (Sunderland, MA: Sinauer, 2003), p. 34.

p. 2 **Others, known as "gonidia":** D.L. Kirk, The ontogeny and phylogeny of cellular differentiation in *Volvox*, *Trends in Genetics* 4(1988):32–36; D.L. Kirk, *Volvox: The Molecular-Genetic Origins of Multicellularity and Cellular Differentiation* (Cambridge: Cambridge University Press, 1998), pp. 33; 115–116.

p. 2 **dormant until next year's spring rains:** Kirk, *Volvox*, pp. 51; 126–127; Gilbert, *Developmental Biology*, pp. 36–38.

p. 2 **the *Oxford Dictionary* definition of a society:** *The Oxford Dictionary and Thesaurus, American Edition* (New York: Oxford University Press, 1996).

p. 3 **"some coordination, some integration or communication":** J.T. Bonner, *Cells and Societies* (Princeton: Princeton University Press, 1955), p. 4.

p. 3 **the gonidia, the reproductive cells:** A. Hallman, K. Godl, and M. Sumper, The highly efficient sex-inducing pheromone system of *Volvox*, *Trends in Microbiology* 6(1998):185–189.

p. 5 **a structure known today as the cell nucleus:** A comprehensive account of the discovery of the cell and the evolution of the "cell doctrine"—the idea that living organisms are composed of cells—one of the central tenets of biology, can be found in H. Harris, *The Birth of the Cell* (New Haven, CT: Yale University Press, 1999).

1
SMALL TALK

p. 10 **textbooks refer to as the "random walk":** H.C. Berg, *Random Walks in Biology* (Princeton, NJ: Princeton University Press, 1983).

p. 10 **"before opening your eyes becomes irresistible?":** R. Llinas, *I of the Vortex: From Neurons to Self* (Cambridge, MA: The MIT Press, 2001), p. 18.

p. 14 **its five senses:** F.W. Dahlquist, Amplification of signaling events in bacteria, *Science's STKE* 2002 (2002), online at www.stke.org/cgi/content/full/OC_sigtrans;2002/132/ pe24; R.B. Bourret and A.M. Stock, Molecular information processing: Lessons from bacterial chemotaxis, *Journal of Biological Chemistry* 277 (2002):9625–9628. The Science Signal Transduction Knowledge Environment (STKE; www.stke.org) is a must-see resource for anyone interested in cell signaling research. In addition to an electronic publication featuring original articles as well as reviews on a wide range of topics relevant to signaling research, the STKE also offers a library of "Connections Maps" illustrating the components of critical signaling pathways and their interrelationships, a directory of signaling researchers, a guide to upcoming meetings devoted to signaling, and full-text access to relevant articles from the medical literature. Access to the entire STKE database requires an individual subscription ($69 per year), but a number of the features are free to anyone who registers at the website. A similar "knowledge portal" devoted to cell signaling is maintained by the journal *Nature* (Nature Signaling Gateway, at www.signaling-gateway.org). More information on the Signaling Gateway can be found in the notes for Chapter 6.

p.14 **a pair of identical polypeptide subunits:** S.A. Chervitz and J.J. Falke, Molecular mechanism of transmembrane signaling by the aspartate receptor: A model, *Proceedings of the National Academy of Sciences* 93(1996):2545–2550.

p. 16 **add the nuance of context:** Bourret and Stock (2002).

p. 17 **the tails of chemotactic receptors:** Dahlquist (2002); Bourret and Stock (2002).

p. 17 **over five orders of magnitude:** A.M. Stock, A nonlinear stimulus-response relation in bacterial chemotaxis, *Proceedings of the National Academy of Sciences* 96(1999):10945–10947.

p. 17 **a primitive form of memory:** J.B. Stock, M.N. Levit, and P.M. Wolanin, Information processing in bacterial chemotaxis, *Science's STKE* 2002 (2002), http://www.stke.org/cgi/content/full/OC_sigtrans;2002/132/pe25.

p. 18 **"The modern era begins":** J. Barzun, *From Dawn to Decadence: 500 Years of Western Cultural Life* (New York: HarperCollins, 2000), p. 3.

p. 20 **culture the beach along with the bacteria:** T. Kaerberlein, K. Lewis, and S. Epstein, Isolating "uncultivable" microorganisms in pure culture in a simulated environment, *Science* 296(2002):1127–1129.

p. 22 **squid avoids becoming someone else's:** E.G. Ruby, Lessons from a cooperative bacterial-animal association: The *Vibrio fischeri-Euprymna scolopes* light organ symbiosis, *Annual Review of Microbiology* 50(1996):591–624. A recent study suggests that the bacteria not only populate their safe haven, but also engineer it: a bacterial peptide, tracheal cytotoxin, released by *V. fischeri* promotes the development of the light organ. See T.A. Koropatnick, J.T. Engle, M.A. Apicella, et al., Microbial factor-mediated development in a host-bacterial mutualism, *Science* 306(2004):1186–1188.

p. 22 **"*N*-(3-oxohexanoyl)-homoserine lactone":** C. Fuqua, M.R. Parsek, and E.P. Greenberg, Regulation of gene expression by cell-to-cell communication: Acyl-homoserine lactone quorum sensing, *Annual Review of Genetics* 35(2001):439–468.

p. 22 **the expression of the luciferase gene:** K.H. Nealson, T. Platt, and J.W. Hastings, Cellular control of the synthesis and activity of the bacterial luminescent system, *Journal of Bacteriology* 104(1970):313–322.

p. 23 **both receptor and response regulator:** S. Schauder and B.L. Bassler, The languages of bacteria, *Genes & Development* 15(2001):1468–1480; M.B. Miller and B.L. Bassler, Quorum sensing in bacteria, *Annual Review of Microbiology* 55(2001): 165–199.

p. 25 **acyl-HSLs for short:** Miller and Bassler (2001); S.C. Winans and B.L. Bassler, Mob psychology, *Journal of Bacteriology* 184(2002):873–883.

p. 25 **controlling critical genes:** Schauder and Bassler (2001); Miller and Bassler (2001); Winans and Bassler (2002).

p. 25 **Gram-positive bacteria . . . also count heads:** Schauder and Bassler (2001); Miller and Bassler (2001); Winans and Bassler (2002).

p. 27 **"the development of multicellular organisms":** Winans and Bassler (2002).

p. 27 **"one language and few words":** Gen. 11:1 Revised Standard Version.

p. 27 **and undermines diplomacy:** B.Wuethrich, Learning the world's languages before they vanish, *Science* 288(2000):1156–1159.

p. 27 **a Polish doctor, Ludwig L. Zamenhof:** The Esperanto League for North America Homepage, The ELNA/Esperanto FAQ List, www.esperanto-usa.org/elnafaq.html.

p. 27 **and 20 times easier than Chinese:** Esperanto Homepage; Esperanto: FAQs, The Virtual Esperanto Library, www.esperanto.net/veb/faq-2.html.

p. 28 **perhaps 2 million speakers worldwide:** Virtual Esperanto Library.

p. 29 **the causative agent in bubonic plague:** Schauder and Bassler (2001); Miller and Bassler (2001); Winans and Bassler (2002).

p. 30 **an unusual bolt: a boron atom:** X. Chen, S. Schauder, N. Potier, et al., Structural identification of a bacterial quorum-sensing signal containing boron, *Nature* 415(2002):545–549.

p. 32 **call a "biofilm":** J.W. Costerton, Z. Lewandowski, D.E. Caldwell, et al., Microbial biofilms, *Annual Review of Microbiology* 49(1995):711–745; G. O'Toole, H.B. Kaplan, and R. Kolter, Biofilm formation as microbial development, *Annual Review of Microbiology* 54(2000):49–79.

p. 33 **draining away dissolved waste products:** Single-species biofilms generated in the laboratory are often just one cell thick. But in natural settings, biofilm thickness varies widely, depending on the number of species contained in the biofilm and the stage of formation, among other factors.

p. 33 **"Chefs and grocers may settle":** P. Watnick and R. Kolter, Biofilm, city of microbes, *Journal of Bacteriology* 182(2000):2675–2679.

p. 34 **to excavate its interconnected canals:** D.G. Davies, M.R. Parsek, J.P.

Pearson, et al., The involvement of cell-to-cell signals in the development of a bacterial biofilm, *Science* 280(1998):295–298; M. Whitely, K.M. Lee, and E.P. Greenberg, Identification of genes controlled by quorum sensing in *Pseudomonas aeruginosa, Proceedings of the National Academy of Sciences* 96(1999):13904–13909; C. Fuqua and E.P. Greenberg, Listening in on bacteria: Acyl-homoserine lactone signaling, *Nature Reviews Molecular Cell Biology* 3(2002):685–695.

p. 34 **architectural detail or social convention:** Davies, et al. (1998); M. Chicurel, Slimebusters, *Nature* 408(2000):284–286.

p. 34 **and recruit passersby:** S. Schauder and B.L. Bassler (2001); Chicurel (2000).

p. 34 **solid foundation is an anchor in a storm:** Another advantage for the bacteria (but not their host) is that living in a biofilm seems to confer resistance to antibiotics. The reasons behind this invincibility may be architectural—the biofilm structure impedes the penetration of the antibiotic or reduces its efficacy because of regional variations in nutrient availability or waste concentration—or physiological—within the biofilm, bacteria may divide more slowly or differentiate into a drug-insensitive state. Biofilm-mediated antibiotic resistance is thought to play a critical role in the persistence of chronic infections, such the recurrent lung infections characteristic of cystic fibrosis. See: J.W. Costerton, P.S. Stewart, and E.P. Greenberg, Bacterial biofilms: A common cause of persistent infections, *Science* 284(1999):1318–1322; and B. Schachter, Slimy business—The biotechnology of biofilms, *Nature Biotechnology* 21(2003):361–366.

p. 34 **"If the bacteria were unable to escape":** Watnick and Kolter (2000).

p. 35 **"deeply rooted in the nature of things":** C. Alexander, S. Ishikawa, and M. Silverstein with M. Jacobson, I. Fiksdahl-King, S. Angel, *A Pattern Language: Towns, Buildings, Construction* (New York: Oxford University Press, 1977), p. xvii.

p. 35 **"In a gothic cathedral":** C. Alexander, *The Timeless Way of Building* (New York: Oxford University Press, 1979), p. 86.

p. 35 **"each sidewalk . . . includes":** Ibid., p. 73.

p. 36 **a problem that "occurs over and over again":** Alexander, et al., *A Pattern Language,* p. x.

p. 36 **"a deep and inescapable property":** Ibid., p. xiv.

p. 36 **"patterns which specify connections between patterns":** Alexander, *The Timeless Way*, p. 187.

p. 36 **"features that bear a particular relationship to each other":** Ibid., p. 88.

p. 39 **the sophisticated biological sentences of higher organisms:** "Conservation," in this case, refers to the preservation of mechanisms; the actual signaling molecules used by bacteria and animal cells differ.

2
BUILD IT AND THEY WILL TALK

p. 42 **significant improvement in energy efficiency:** B. Alberts, A. Johnson, J. Lewis, M. Raff, K. Roberts, and P. Walker, *Molecular Biology of the Cell*, 4th ed. (New York: Garland Science, 2002), pp. 824–826.

p. 42 **"it was the horse's value as a mount":** S. Budiansky, *The Nature of Horses: Exploring Equine Evolution, Intelligence, and Behavior* (New York: The Free Press, 1997), p. 55.

p. 42 **we know today as mitochondria:** Alberts, et al. *Molecular Biology of the Cell,* pp. 29–33; M.W. Gray, G. Burger, and B.F. Lang, Mitochondrial evolution, *Science 283*(1999):1476–1481.

p. 43 **permitted them to move freely about:** Alberts, et al., *Molecular Biology of the Cell,* pp. 29–33; J Gerhart and M. Kirschner, *Cells, Embryos, and Evolution: Toward a Cellular and Developmental Understanding of Phenotypic Variation and Evolutionary Adaptability* (Malden, MA: Blackwell Science, 1997), pp. 9–15.

p. 43 **forswore rigid walls that prevented real intimacy for good:** Gerhart and Kirschner, *Cells, Embryos, and Evolution,* pp. 9–15.

p. 46 **"the one realm . . . available to escape competition":** J.T. Bonner, *First Signals: The Evolution of Multicellular Development* (Princeton, NJ: Princeton University Press, 2000): pp. 51–52.

p. 46 **"better coordination of the adhering cells":** Bonner, *First Signals,* p. 7.

p. 46 **life's first true tissue:** Gerhart and Kirschner, *Cells, Embryos, and Evolution,* p. 244.

p. 48 **black bile, dark and inert as earth:** S. Finger, *Minds Behind the Brain: A History of the Pioneers and Their Discoveries* (New York: Oxford University Press, 2000), p. 31.

p. 49 **"each gland . . . is the workshop":** A.Q. Maisel, *The Hormone Quest* (New York: Random House, 1965), p. 7.

pp. 49–50 **"A diseased condition" of the adrenal glands:** Addison's original description of the disease which bears his name, quoted here, is reproduced in R. LeBaron, *Hormones: A Delicate Balance* (New York: Pegasus, 1972), pp. 65–66.

p. 50 **a stranger with a vial . . . the boy's radial artery:** H.M. Leicester, *Development of Biochemical Concepts from Ancient to Modern Times* (Cambridge, MA: Harvard University Press, 1974), pp. 226–227; "Sharpey-Schafer, Edward Albert (born Schäfer)," http://www.cartage.org.lb/en/themes/Biographies/MainBiographies/S/Sharpey-Schafer/l; Finger, *Minds Behind the Brain,* p. 261; B.D. Gomperts, P.E.R. Tatham, and I.M. Kramer, *Signal Transduction* (San Diego: Academic Press, 2002), pp. 8–9. Gomperts notes that accounts of the incident differ somewhat, with Schäfer's own description rather more clinical than the exuberant recollections of some of his colleagues.

p. 51 **"So, Professor Schafer makes the injection . . .":** J.C. Krantz, Jr., John J. Abel and Epinephrine—The First Hormone, in *Historical Medical Classics Involving New Drugs* (Baltimore: Williams & Wilkins, 1974), pp. 42–48.

p. 51 **"adrenaline" after the gland that secreted it:** Krantz, et al. (1974); J.K. Aronson, "Where name and image meet"—The argument for "adrenaline," *British Medical Journal* 320(2000):506–509. The substance isolated by Abel was, in fact, not the pure hormone, but the benzoyl derivative (the benzoyl moiety was accidentally introduced during the extraction procedure). Takamine learned the process while a visiting scientist in Abel's laboratory; upon returning to his own laboratory, he treated the final product with dilute ammonia to remove the impurity. Takamine patented his substance and subsequently transferred the rights to the pharmaceutical firm Parke,

Davis & Co., which marketed it under the trade name "Adrenalin," to denote its origin. Although Abel's term "epinephrine" is preferred in the United States, historically, "adrenaline," as Aronson notes, is probably more correct. Along the same lines, the chemical precursor of adrenaline—a signal in its own right—can be called either "norepinephrine" or "noradrenaline."

Takamine was a gracious ambassador as well as a skilled chemist. Upon learning of President Taft's wife's infatuation with cherry trees, he conspired with the mayor of Tokyo to arrange a gift of 3,000 Japanese cherry trees. Planted along the Tidal Basin, they became the inspiration for Washington's annual Cherry Blossom Festival.

p. 53 **the names of the bones:** L.E. Limbird, *Cell Surface Receptors*. 1. Historical perspective (Boston: Martinus Nijhoff, 1986), pp. 3–6; J.C. Krantz, Jr., Paul Ehrlich and the magic bullet arsphenamine, in *Historical Medical Classics Involving New Drugs* (Baltimore: Williams & Wilkins, 1974), pp. 51–57; P. DeKruif, *The Microbe Hunters* (New York: Harvest Books, 1966), pp. 334–358.

p. 53 **the recipient of dozens of honorary awards:** Gomperts et al., *Signal Transduction*, pp. 15–16; Limbird, *Receptors*, pp. 3, 5–6.

p. 53 **explanation for such specificity:** Limbird, *Receptors*, p. 3.

p. 54 **both vied for this same target:** Limbird, *Receptors*, p. 5.

p. 54 **"*Corpora non agunt nisi fixata*":** quoted in Limbird, *Receptors*, p. 5.

pp. 54–55 **able to deconstruct the polymer:** Glycogen is also stored in the muscles; this resource is the first tapped. But when muscle glycogen is exhausted by protracted effort, the body turns to the glycogen stored in the liver for sustenance until the crisis has passed.

p. 55 **the alarm clock that actually rouses glycogen phosphorylase:** J.A. Beavo and L.L. Brunton, Cyclic nucleotide research—Still expanding after half a century, *Nature Reviews Molecular Cell Biology* 3(2002):710–717; G.A. Robison, R.W. Butcher, and E.W. Sutherland, *cAMP* (New York: Academic Press, 1971); E. Krebs and E. Fischer, The phosphorylase B to A converting enzyme of rabbit skeletal muscle, *Biochimica Biophysica Acta* 1989(1956):302–309.

p. 55 **as robust as that found in a living cell:** Robison, et al. (1971); J. Berthet, T.W. Rall, and E.W. Sutherland, The relationship of epinephrine and glucagon to liver phosphorylase. IV. Effect of epinephrine and glucagon on the reactivation of phosphorylase in liver homogenates, *Journal of Biological Chemistry* 224(1957):463–475.

p. 56 **"cyclic AMP" for short:** E.W. Sutherland and T.W. Rall, Fractionation and characterization of a cyclic adenine ribonucleotide formed by tissue particles, *Journal of Biological Chemistry* 232(1958):1077–1091.

p. 57 **a paper in the prestigious journal:** R.J. Lefkowitz, J. Roth, W. Pricer, et al., ACTH receptors in the adrenal: Specific binding of ACTH-^{125}I and its relation to adenyl cyclase, *Proceedings of the National Academy of Sciences* 65(1970):745–752.

p. 58 **"without seriously hurting the organism":** Krantz (1974).

p. 58 **membranes from heart and brain:** R.J. Lefkowitz, Identification and regulation of alpha- and beta-adrenergic receptors, *Federation Proceedings* 37(1978):123–129; M.G. Caron and R.J. Lefkowitz, Catecholamine receptors: Structure, function, and regulation, *Recent Progress in Hormone Research* 48(1993):277–319; R.J. Lefkowitz, M.G. Caron, and G.L. Stiles, Mechanisms of membrane-receptor regulation. Biochemical, physiological, and clinical insights derived from studies of the adrenergic receptors, *New England Journal of Medicine* 310(1984):1570–1579.

p. 59 **adrenergic receptors . . . had multiple personalities:** R. P. Ahlquist, A study of the adrenergic receptors, *American Journal of Physiology 153*(1948):568–586.

p. 59 **α receptors into two groups . . . β receptors into three:** Caron and Lefkowitz (1993); R.J. Lefkowitz, A. De Lean, B.B. Hoffman, et al., Molecular pharmacology of adenylate cyclase-coupled α- and β-adrenergic receptors, *Advances in Cyclic Nucleotide Research* 14(1981):145–161.

p. 59 **had corralled the β_2-adrenertgic receptor:** R.G.L. Shorr, R.J. Lefkowitz, and M.G. Caron, Purification of the beta-adrenergic receptor. Identification of the hormone binding subunit, *Journal of Biological Chemistry* 256(1981):5820–5826.

p. 59 **proof that receptor subtypes were actually distinct proteins:** R.G.L. Shorr, M.W. Strohsacker, T.N. Lavin, et al., The beta 1-adrenergic receptor of the turkey erythrocyte. Molecular heterogeneity revealed by purification and photoaffinity labeling, *Journal of Biological Chemistry* 257(1982):12341–12350; R.J. Lefkowitz, The superfamily of heptahelical receptors, *Nature Cell Biology* 2(2000):E133–E136; R.A.F. Dixon, B.K. Kobilka, D.J. Strader, et al., Cloning of the gene and cDNA for mammalian beta-adrenergic receptor and homology with rhodopsin, *Nature* 321(1986): 75–79.

p. 60 **watery environment on either side:** Lefokowitz (2000); Dixon, et al. (1986); H.G. Dohlman, J. Thorner, and M.G. Caron, et al. Model systems for the study of seven-transmembrane-segment receptors, *Annual Review of Biochemistry* 60(1991):653–688.

p. 60 **the most important ingredient in your standard cell membrane:** For the information on the life and work of Martin Rodbell, I am especially indebted to "The Martin Rodbell Collection" in Profiles in Science, a website maintained by the National Library of Medicine at http://profiles.nlm.nih.gov/GG/Views/Exhibit/narrative/cells.html.

p. 60 **"is in essence a communication device":** Rodbell Collection.

p. 61 **which had been christened "adenylyl cyclase," was the amplifier:** M. Rodbell, Signal transduction: Evolution of an idea, Nobel lecture, presented December 8, 1994.

p. 61 **carried out under the same conditions:** Rodbell, ibid.; M. Rodbell, H.M.J. Krans, S.L. Pohl, et al., The glucagon-sensitive adenyl cyclase system in plasma membranes of rat liver: Effects of guanylnucleotides on binding of ^{125}I-glucagon, *Journal of Biological Chemistry* 246(1971):1872–1876.

p. 61 **"I tested many [other] types . . . of nucleotides":** Rodbell, ibid.

p. 61 **it quit making cyclic AMP:** M. Rodbell, L. Birnbaumer, S. Pohl, et al., The glucagon-sensitive adenyl cyclase system in plasma membranes of rat liver: An obligatory role of guanylnucleotides in glucagon action, *Journal of Biological Chemistry* 246(1971):1877–1882.

p. 62 **distinct from the binding site for glucagon:** Y. Salomon and M. Rodbell, Evidence for specific binding sites for guanine nucleotides in adipocyte and hepatocyte plasma membranes: A difference in fate of GTP and guanosine 5′-(beta, gamma-imino) triphosphate, *Journal of Biological Chemistry* 250(1975):7245–7250.

p. 62 **turn GTP into guanosine diphosphate, GDP:** D. Cassel, H. Levkovitz, and Z. Selinger, The regulatory GTPase cycle of turkey erythrocyte adenylate cyclase, *Journal of Cyclic Nucleotide Research* 3(1977):393–406; D. Cassel and Z. Selinger, Catecholamine-stimulated GTPase activity in turkey erythrocyte membranes, *Biochimica Biophysica Acta* 452(1976):538–551.

p. 62 **an extended family of such G proteins:** A.G. Gilman, G proteins: Transducers of receptor-generated signals, *Annual Review of Biochemistry* 56(1987):615–649; Gomperts, et al., *Signal Transduction*, pp. 71–105.

p. 63 **"the only person who was ever named after a textbook":** "Alfred G. Gilman—Autobiography," Nobel e-Museum Website, www.nobel.se/medicine/laureates/1994/gilman-autobio.html.

p. 63 **fate seemed to have hitched his star to adenylyl cyclase:** Ibid.; A.G. Gilman, G proteins and regulation of adenylyl cyclase, Nobel lecture, delivered December 8, 1994.

p. 63 **they died in their dishes:** V. Daniel, G. Litwack, and G.M. Tomkins, Induction of cytolysis of cultured lymphoma cells by adenosine 3′,5′-cyclic monophosphate and the isolation of resistant variants, *Proceedings of the National Academy of Sciences* 70(1973):76–79.

p. 64 **the enzyme was fine:** P.A. Insel, M.E. Maguire, A.G. Gilman, et al., Beta adrenergic receptors and adenylate cyclase: Products of separate genes? *Molecular Pharmacology* 12(1976):1062–1069; T. Haga, E.M. Ross, and H.J. Anderson, et al., Adenylate cyclase permanently uncoupled from hormone receptors in a novel variant of S49 mouse lymphoma cells, *Proceedings of the National Academy of Sciences U.S.A.* 74(1977):2016–2020; E.M. Ross and A.G. Gilman, Resolution of some components of adenylate cyclase necessary for catalytic activity, *Journal of Biological Chemistry* 252(1977):6966–6969.

p. 64 **plenty of fully functional adrenergic receptors:** Insel, et al. (1976); Haga, et al. (1977); Ross and Gilman (1977).

p. 64 **a confederacy of three polypeptides:** Gilman (1987); Gomperts, et al., *Signal Transduction*, pp. 71–105; J.K. Northup, P.C. Sternweis, M.D. Smigel, et al., Purification of the regulatory component of adenylate cyclase, *Proceedings of the National Academy of Sciences U.S.A.* 77(1980):6516–6520; P.C. Sternweis, J.K. Northup, M.D. Smigel, et al., The regulatory component of adenylate cyclase. Purification and properties, *Journal of Biological Chemistry* 256(1981):11517–11526.

p. 65 **X-ray diffraction patterns and magnetic resonance spectra:** D.G. Lambright, J. Sondek, A, Bohm, et al., The 2.0 A structure of a heterotrimeric G protein, *Nature* 379(1996):311–319; J. Sondek, A. Bohm, D.G. Lambright, et al., Crystal structure of a G-protein beta gamma dimer at 2.1A resolution, *Nature* 379 (1996):369–374; M. Wall, et al., The structure of the G protein heterotrimer Gi alpha 1 beta 1 gamma 2, *Cell* 83(1995):1047–1058; D.E. Clapham, The G-protein nanomachine, *Nature* 379(1996):297–299.

p. 65 **"Biological communication" . . . "consists of a complex meshwork":** Rodbell, Nobel lecture (1994).

p. 66 **the same seven-helix, membrane-spanning structure:** K.L. Pierce, R.T. Premont, and R.J. Lefkowitz, Seven-transmembrane receptors, *Nature Reviews Molecular Cell Biology* 3(2002):639–650. Maps of representative G-protein pathways can be found online at the *Science* Signal Transduction Knowledge Environment (STKE) Website at www.stke.org.

p. 66 **the G-protein-coupled light receptor, rhodopsin, were look-alikes:** Dixon, et al. (1986).

p. 67 **than with any other introductory word or phrase:** Alberts et al., *Molecular Biology of the Cell*, p. 852.

p. 69 **then works with the others to complete the job:** Ibid., pp. 140–141; 145–146.

p. 69 **Domains . . . may have started off as proteins themselves:** Ibid.

p. 69 **a "consensus sequence" on another protein:** T. Pawson and P. Nash, Assembly of cell regulatory systems through protein interaction domains, *Science* 300 (2003):445–452.

p. 69 **the pool of potential partners:** Ibid.

p. 70 **"expanding the possibilities for combinatorial association":** M. Ptashne and A. Gann, *Genes and Signals* (Cold Spring Harbor, NY: Cold Spring Harbor Laboratory Press, 2002), p. xvi.

p. 70 **the mutation that transformed c-*fps* into v-*fps*:** I. Sadowski, J.C. Stone, and T. Pawson, A noncatalytic domain conserved among cytoplasmic protein-tyrosine kinases modifies the kinase function and transforming activity of Fujinami sarcoma virus $P130^{gag\text{-}fps}$, *Molecular and Cellular Biology* 6(1986):4396–4408.

p. 71 **including Src:** Ibid.

p. 71 **blocking access to the enzyme's active site:** S.R. Hubbard, M. Mohammadi, and J. Schlessinger, Autoregulatory mechanisms in protein-tyrosine kinases, *Journal of Biological Chemistry* 273(1998):11987–11990.

p. 71 **"SH2," for "Src-homology 2":** Sadowski, et al. (1986); T. Pawson, G.D. Gish, and P. Nash, SH2 domains, interaction modules, and cellular wiring, *Trends in Cell Biology* 11(2001):504–511; T. Pawson, Protein modules and signaling networks, *Nature* 373(1995):573–580.

p. 71 **catalyzed by a trio of kinases:** R. Triesman. Regulation of transcription by MAP kinase cascades, *Current Opinion in Cell Biology* 8(1996):205–215; C. Widmann, S. Gibson, M.B. Jarpe, et al., Mitogen-activated protein kinase: Conservation of a three-kinase module from yeast to human, *Physiological Reviews* 79 (1999): 143–180; L. Chang and M. Karin, Mammalian MAP kinase signaling cascades, *Nature* 410(2001):37–40.

p. 72 **a GTPase known as Ras:** Pawson, et al. (2001); Widmann, et al. (1999).

p. 72 **by *src*-homology interaction domains:** Pawson, et al. (2001); Pawson (1995); J.D. Scott and T. Pawson, Cell communication: The inside story, *Scientific American*, June 2000:72–79.

p. 74 **in retaliation for their expulsion by $G\alpha$:** Pierce et al. (2002); S.J. Perry and R. J. Lefkowitz, Arresting developments in heptahelical receptor signaling and regulation, *Trends in Cell Biology* 12(2002):130–138.

p. 74 **a prelude to exile or even execution:** Ibid.

p. 74 **turns them over to the care of another G protein:** Ibid.

p. 75 **dressed up as an adaptor:** Ibid.

p. 75 **"Patterns" . . . "need the context of others to make sense":** C. Alexander, *The Timeless Way of Building* (New York: Oxford University Press, 1979), p. 312.

pp. 75–76 **Begin your building project . . . and maintain an optimal population density:** C. Alexander, S. Ishikawa, and M. Silverstein with M. Jacobson, I. Fiksdahl-King, S. Angel, *A Pattern Language: Towns, Buildings, Construction* (New York: Oxford University Press, 1977), pp. 21–25; 70–74; 304–309; 468–472; 508–512; 614–617; 828–832; 893–899; 1006–1008; 1018–1022; 1100–1104.

p. 76 **"helps to complete those larger patterns":** Ibid., p. xii.

p. 76 **"where we most expect to find variation, we find conservation":** Gerhart and Kirschner, *Cells, Embryos and Evolution*, p. 1.

p. 77 **hand in hand with the evolution of these organisms:** Ibid., pp. 45–89.

p. 78 **Sixteen different genes . . . encoded by five. . . . Twelve genes specify γ subunits:** Gomperts, et al., *Signal Transduction*, pp. 79–82.

p. 79 **to the diversity of these organisms:** Gerhart and Kirschner, *Cells, Embryos, and Evolution*, pp. 45–89.

p. 79: **"Patterns have enormous power and depth":** Alexander, *The Timeless Way*, p. 98.

p. 79 **"a few thousand gene products":** Specificity in signal transduction, Samuel Lunenfeld Research Institute website, www.mshri.on.ca/pawson/research2.html.

3
PLAITING THE NET

p. 81 **the apical ectodermal ridge:** J.W. Saunders Jr. The proximo-distal sequence of the origin of the parts of the chick wing and the role of the ectoderm, *Journal of Experimental Zoology* 108(1948):363–403.

p. 81 **at right angles to its normal orientation:** Saunders (1948); D.A. Summerbell, A quantitative analysis of the effect of excision of the AER from the chick limb bud, *Journal of Embryology and Experimental Morphology* 32(1974):651–660; D.A. Rowe and J.F. Fallon, The proximodistal determination of skeletal parts in the developing chick leg, *Journal of Embryology and Experimental Morphology* 68(1982):1–7; E. Zwilling, Interaction between limb bud ectoderm and mesoderm in the chick embryo. I. Axis establishment, *Journal of Experimental Zoology* 132(1956): 157–171.

p. 82 **indistinguishable from a normal limb:** G. R. Martin, The roles of FGFs in the early development of vertebrate limbs, *Genes & Development* 12(1998):1571–1585; L. Niswander, C. Tickle, A. Vogel, et al., FGF-4 replaces the apical ectodermal ridge and directs outgrowth and patterning of the limb, *Cell* 75(1993):579–587; J.F. Fallon, M.A. López, M.P. Ros, et al., FGF-2: Apical ectodermal ridge growth signal for chick limb development, *Science* 264(1994):104–107.

p. 82 **in the space between wing and leg:** Martin (1998); M.J. Cohn, J.-C. Izpisúa-Belmonte, H. Abud, et al., Fibroblast growth factors induce additional limb development from the flank of chick embryos, *Cell* 80 (1995):739–746; A. Vogel, C. Rodriguez, and J.-C. Izpisúa-Belmonte, Involvement of FGF-8 in initiation, outgrowth and patterning of the vertebrate limb, *Development* 122(1996):1737–1750.

p. 83 **cells can be substituted for the bead:** H. Ohuchi, T. Nakagawa, M.

Yamauchi, et al., An additional limb can be induced from the flank of the chick embryo by FGF-4, *Biochemical and Biophysical Research Communications* 204(1995): 809–816; H. Ohuchi, T. Nakagawa, A. Yamamoto, et al., The mesenchymal factor, FGF10, initiates and maintains the outgrowth of the chick limb bud through interaction with FGF8, an apical ectodermal factor, *Development* 124(1997):2235–2244.

p. 83 **in mammals as it is in birds:** H. Min, D.M. Danilenko, S.A. Scully, et al., Fgf-10 is required for both limb and lung development and exhibits striking functional similarity to *Drosophila* branchless, *Genes and Development* 12(1998):3156–3161; K. Sekine, et al., Fgf10 is essential for limb and lung formation, *Nature Genetics* 21(1999):138–141.

p. 83 **instruct the optic vesicle to make a retina:** Y. Furata and B.L.M. Hogan, BMP4 is essential for lens induction in the mouse embryo, *Genes and Development* 12(1998):3764–3775; A. Vogel-Höpker,T. Momose, H. Rohrer, et al., Multiple functions of fibroblast growth factor-8 (FGF-8) in chick eye development, *Mechanisms of Development* 94(2000):25–36.

p. 83 **given you a duplicate set of fingers:** A. López-Martinez, et al., Limb-patterning activity and restricted posterior localization of the amino-terminal product of sonic hedgehog cleavage, *Current Biology* 5(1995):791–796.

p. 84 **"this is normally impossible in architecture":** J.J. Coulton, *Ancient Greek Architects at Work* (Ithaca, NY: Cornell University Press, 1997), p. 53.

p. 84 **"to communicate the architect's intention to the builders":** Ibid.

p. 84 **a technique to specify proportions:** Ibid., pp. 54–55.

p. 84 **"*Ex ovo omnia*":** C. Singer, *A History of Biology to About the Year 1900: A General Introduction of the Study of Living Things*, 3rd ed. (London: Abelard-Schuman, 1962), p. 466.

p. 85 **"as we read in the poems of Orpheus":** Quoted in G.W. Greg, The organizer, *Scientific American* 197(1957):79–88.

p. 85 **"If organic form is not original":** W. Coleman, *Biology in the 19th Century: Problems of Form, Function, and Transformation* (New York: Wiley, 1971), p. 42.

p. 85 **the movement of the stars or the actions of gravity:** Singer, *History of Biology*, pp. 465–467; L.N. Magner, *A History of the Life Sciences*, 2nd ed. (New York: Marcel Dekker, 1994), pp. 171–186; S.F. Gilbert, *Developmental Biology*, 7th ed. (Sunderland, MA: Sinauer, 2003), pp. 6–7.

p. 86 **when they peered into their microscopes:** Singer, *History of Biology*, pp. 506–507; Magner, *History of the Life Sciences*, pp. 176–178.

p. 86 **one of the first of these revisionists:** Singer, *History of Biology*, pp. 469–471; Coleman, p. 41.

p. 86 **how the tadpole got its eyes:** S.F. Gilbert, A selective history of induction II: Spemann's induction experiments, online at http://www.devbio.com.

p. 87 **developed on that side of the embryo:** Gilbert, Selective history of induction; C. Fulton and A. O. Klein, *Explorations in Developmental Biology* (Cambridge, MA: Harvard University Press, 1976), pp. 284–287.

p. 87 **"by which the genes of a zygote" . . . "morphogens":** A.M. Turing, The chemical basis of morphogenesis, *Philosophical Transactions of the Royal Society of London B* 327(1952):37–72.

p. 88 **might translate into biological structure:** Turing (1952); Gilbert, *Developmental Biology*, pp. 20–21; P. Ball, *The Self-made Tapestry* (New York: Oxford University Press, 1999), pp. 78–83.

p. 88 **"a technique of design":** Coulton, *Ancient Greek Architects*, p. 53.

p. 88 **in the early rounds of cell division:** Gilbert, *Developmental Biology*, pp. 59–60.

p. 89 **that biologists call "transcription factors":** Comprehensive summaries of eukaryotic gene expression can be found in several leading textbooks; see, for example B. Alberts, A. Johnson, J. Lewis, et al., *Molecular Biology of the Cell*, 4th ed. (New York: Garland Science, 2002), pp. 379–415; or T.D. Pollard and W.C. Earnshaw, *Cell Biology* (Philadelphia: Saunders), pp. 215–237.

p. 90 **use the information stored in its genome selectively:** F. Jacob and J. Monod, Gene regulatory mechanisms in the synthesis of proteins, *Journal of Molecular Biology* 3(1961):318–356.

p. 92 **"A man modeling a clay figure":** Coulton, *Ancient Greek Architects*, p. 53.

p. 93 **"ensure that the lower parts of the building":** Ibid.

p. 93 **a "kind of living developmental fossil":** J.T. Bonner, *First Signals: The Evolution of Multicellular Development* (Princeton, NJ: Princeton University Press, 2000), p. 74.

p. 94 **"a directional arrow that points down the slope":** Ball, *Self-Made Tapestry*, p. 100.

p. 95 **a replica of the French flag:** L. Wolpert, R. Beddington, J. Brockes, et al., *Principles of Development* (Oxford: Oxford University Press, 1998), pp. 19–20.

p. 95 **one of its most significant challenges:** In addition to the technical challenges posed by the isolation and identification of morphogens themselves, it is increasingly clear that gradient formation in the living embryo is more complicated than prototypes like the "French flag model" might suggest. Passive diffusion, for example, is not the only way to create a gradient; an ambitious organism—or a morphogen-maker's meddling neighbors—can hand-carry the morphogen from cell to cell or bind and impede its movement to shape local concentrations in highly precise ways. Moreover, cells exposed to the morphogen can only respond after they perceive and interpret the signal. As a result, the distribution and affinity of morphogen receptors or other signaling elements can also influence the final pattern set up by the morphogen. See J.B. Gurdon and P.Y. Bourillot, Morphogen gradient interpretation, *Nature* 413(2001):797–803, or A.A. Teleman, M. Strigini, and S.M. Cohen, Shaping morphogen gradients, *Cell* 105(2001):559–562 for more information.

p. 96 **"'Death' is a difficult phenotype":** S.F. Gilbert, Christiane Nüsslein-Volhard and *Drosophila* embryogenesis, online at www.devbio.com. As Gilbert notes, conceptual and historical factors also comprised the progress of developmental genetics in this otherwise highly regarded model organism.

p. 96 **rendered their impenetrable shells transparent:** Ibid.

p. 96 **a gene they christened "*bicoid*":** C.N. Nüsslein-Volhard, The bicoid morphogen papers (I): Account from CNV, *Cell* S116(2004):S1–S5.

p. 96 **all-important task of orienting the embryo:** H.G. Frohnhöfer and C. Nüsslein-Volhard, Organization of anterior pattern in *Drosophila* embryo by the maternal gene *bicoid*, *Nature* 324(1986):120–125.

p. 96 **encodes one of these maternal morphogens:** W. Driever and C. Nüsslein-Volhard, A gradient of *bicoid* protein in *Drosophila* embryos, *Cell* 54 (1988):83–93; W. Driever and C. Nüsslein-Volhard, The *bicoid* protein determines position in the *Drosophila* embryo in a concentration-dependent manner, *Cell* 54 (1988):95–104; D. St. Johnston and C. Nüsslein-Volhard, The origin of pattern and polarity in the *Drosophila* embryo, *Cell* 66(1992):201–219.

p. 97 **a gradient that points from tail to head:** St. Johnston and Nüsslein-Volhard (1992); R. Lehmann and C. Nüsslein-Volhard, The maternal gene *nanos* has a central role in posterior pattern formation in the *Drosophila* embryo, *Development* 112(1991):679–691.

p. 97 **more or less evenly throughout:** Gilbert, *Developmental Biology*, p. 272.

p. 97 **Bicoid activates *hunchback*:** W. Driever and C. Nüsslein-Volhard, The Bicoid protein is a positive regulator of *hunchback* transcription in the early *Drosophila* embryo, *Nature* 337(1989):138–143; G. Struhl, K. Struhl, and P.M. Macdonald, The gradient morphogen *bicoid* is a concentration-dependent transcriptional activator, *Cell* 57(1989):1259–1273.

p. 97 **the Caudal protein to the tail end of the embryo:** S.K. Chan and G. Struhl, Sequence-specific RNA binding by Bicoid, *Nature* 388(1997):634; R. Rivera-Pomar, D. Niessling, U. Schmidt-Ott, et al., RNA binding and translational suppression by bicoid, *Nature* 379(1996):746–749; D. Niessing, W. Driever, F. Sprenger, et al., Homeodomain position 54 specifies transcriptional versus translational control by Bicoid, *Molecules and Cells* 5(2000):395–401.

p. 97 **renders it indecipherable:** D.D. Barker, C. Wang, J. Moore, et al., Pumilio is essential for function but not for distribution of the *Drosophila* abdominal determinant, Nanos, *Genes and Development* 6(1992):2312–2326; C. Wreden, A.C. Verrotti, J.A. Schisa, et al., Nanos and pumilio establish embryonic polarity in *Drosophila* by promoting poster deadenylation of *hunchback* mRNA, *Development* 124(1997):3015–3023.

p. 97 **gradient that backs up Nanos:** Gilbert, *Developmental Biology* p. 274.

p. 97 **a mist of Dorsal protein:** St. Johnston and Nüsslein-Volhard (1992); A.J. Courey and J.-D. Huang, The establishment and interpretation of transcription factor gradients in the *Drosophila* embryo, *Biochimica Biophysica Acta* 1261(1995):1–18; S. Roth, D. Stein, and C. Nüsslein-Volhard, A gradient of nuclear localization of the dorsal protein determines dorsoventral pattern in the *Drosophila* embryo, *Cell* 59(1989):1189–1202.

p. 98 **until it's liberated from its minder:** S. Roth, Y. Hiromi, D. Godt, et al., *cactus*, a maternal gene required for the proper formation of the dorsoventral morphogen gradient in *Drosophila* embryos, *Development* 112(1991):371–388.

p. 98 **Dorsal can gambol into nuclei:** Courey and Huang (1995).

p. 98 **from the bottom of the embryo to its back:** Roth, et al. (1989); C.A. Rushlow, K. Han, J.L. Manley, et al., The graded distribution of the *dorsal* morphogen is initiated by selective nuclear transport in *Drosophila* , *Cell* 59(1989):1165–1177; R. Steward, Relocalization of the *dorsal* protein from the cytoplasm to the nucleus correlates with its function, *Cell* 59(1989):1179–1188.

p. 98 **"a coordinate system . . . to specify positions":** Gilbert, *Developmental Biology*, p. 297.

p. 99 **known collectively as the gap genes:** Ibid., p. 279; C. Nüsslein-Volhard and E. Weischaus, Mutations affecting segment number and polarity in *Drosophila*, *Nature* 287(1980):795–801.

p. 99 **expressed in seven alternating stripes:** Gilbert, *Developmental Biology*, pp. 281–283.

p. 99 **segmental pattern of the fly larva:** A. Martinez-Arias and P.A. Lawrence, Parasegments and compartments in the *Drosophila* embryo, *Nature* 313(1985):639–642.

p. 99 **dividing one from the other:** Gilbert, *Developmental Biology*, pp. 285–290; P.W. Ingham, A.M. Taylor, and Y. Nakano, Role of *Drosophila patched* gene in positional signaling, *Nature* 353(1991):184–187; J. Mohler and K. Vani, Molecular organization and embryonic expression of the *hedgehog* gene involved in cell-cell communication in segmental patterning in *Drosophila*, *Development* 115(1992):957–971.

p. 100 **the structures they specify:** Gilbert, *Developmental Biology*, pp. 285–290; E.B. Lewis, A gene complex controlling segmentation in *Drosophila*, *Nature* 276(1978):565–570; K. Harding, C. Wedeen, W. McGinnis, et al., Spatially regulated expression of homeotic genes in *Drosophila*, *Science* 229(1985):1236–1242; M.E. Akam, The molecular basis for metameric patterning in the *Drosophila* embryo, *Development* 101(1987):1–22; J. Gerhart and M. Kirschner, *Cells, Embryos, and Evolution: Toward a Cellular and Developmental Understanding of Phenotypic Variation and Evolutionary Adaptability* (Malden, MA: Blackwell Scientific, 1997), pp. 314–325.

p. 101: **chanting "Decapentaplegic, Decapentaplegic":** C Neumann and S. Cohen, Morphogens and pattern formation, *BioEssays* 19(1997):721–729; R.D. St. Johnston and W.M. Gelbart, *Decapentaplegic* transcripts are localized along the dorsal-ventral axis of the *Drosophila* embryo. *Embo Journal* 6(1987):2785–2791; V.F. Irish and W.M. Gelbart, The *decapentaplegic* gene is required for dorsal-ventral patterning of the *Drosophila* embryo, *Genes & Development* 1(1987):868–879; E.I. Ferguson and K.V. Anderson, Decapentaplegic acts as a morphogen to organize dorsal-ventral pattern in the *Drosophila* embryo, *Cell* 71(1992):451–461; K.A. Wharton et al., An activity gradient of decapentaplegic is necessary for the specification of dorsal pattern elements in the *Drosophila* embryo, *Development* 117(1993):807–822.

p. 101 **their felicitous effects on bone growth:** B.L.M. Hogan, Bone morphogenetic proteins: Multifunctional regulators of vertebrate development, *Genes & Development* 10(1996):1580–1594; Y. Shi and J. Massagué, Mechanisms of TGF-β signaling from cell membrane to the nucleus, *Cell* 113(2003):685–700.

p. 101 **"like an open hand":** Hogan (1996).

p. 101 **have activated the genes *twist* and *snail*:** A. Stathopoulos and M. Levine, Dorsal gradient networks in the *Drosophila* embryo, *Developmental Biology* 246(2002):57–67; R. Steward and S. Govind, Dorsal-ventral polarity in the *Drosophila* embryo, *Current Opinion in Genetics and Development* 3(1993):556–561.

p. 102 **and silences it immediately:** Neumann and Cohen (1997); Hogan (1996); S.D. Podos and E.I. Ferguson, Morphogen gradients: New insights from DPP, *Trends in Genetics* 15(1999):396–402.

p. 104 **an entirely new family of tissues:** Gilbert, *Developmental Biology*, p. 46; Gerhart and Kirschner, *Cells, Embryos, and Evolution*, pp. 351–352.

p. 104 **"greater mobility and larger bodies":** Gilbert, *Developmental Biology*, p. 46.

p. 104 **the three so-called germ layers . . . the endoderm:** Ibid., p. 8; Magner, *History of the Life Sciences*, p. 211. The names "ectoderm," "mesoderm," and "endoderm" for the germ layers were introduced in 1855 by another embryologist, Robert Remak.

p. 104 **"although already destined for different ends":** Quoted in Gilbert, S.F., A selective history of induction: I. Early concepts of embryonic induction, online at www.devbio.com.

p. 105 **"cells required for the formation of specific organs":** Gerhart and Kirschner, *Cells, Embryos, and Evolution,* p. 476.

p. 106 **"The first experiment":** H. Spemann, Nobel lecture, delivered December 12, 1935, online at www.nobel.se/medicine/laureates/1935/spemann-lecture.html.

p. 106 **". . . It became apparent":** Ibid.

p. 106 **an embryo of a light-colored species:** Ibid.; Greg (1957); H. Spemann and H. Mangold, Induction of embryonic primordial by implantation of organizers from a different species, translated from the German by Victor Hamburger and reprinted in Fulton and Klein, *Explorations*, pp. 287–325.

p. 108 **secreted chemical signals, not direct contact:** Greg (1957).

p. 108 **where the sperm pierces the egg:** E.M. DeRobertis, J. Larrain, M. Oelgeschläger, et al., The establishment of Spemann's organizer and patterning of the vertebrate embryo, *Nature Reviews Genetics* 1(2000):171–181.

p. 108 **will be the dorsal side of the embryo:** Ibid.; J.R. Miller, B.A. Rowning, C.A. Larabell, et al., Establishment of the dorsal-ventral axis in *Xenopus* embryos coincides with the dorsal enrichment of Dishevelled that is dependent on cortical rotation, *Journal of Cell Biology* 146(1999):427–437.

p. 109 **"*X*enopus . . . *N*odal-*r*elated" proteins:** DeRobertis, et al. (2000); A.F. Schier and M.M. Shen, Nodal signaling in vertebrate development, *Nature* 403 (2000):385–389. Developmental biologists know the signaling center based in the endoderm as the "Nieuwkoop center," after Pieter Nieuwkoop, who, along with Osamu Nakamura, demonstrated its critical role in the induction of the mesoderm and the formation of the organizer. See, for example, P.D. Nieuwkoop, The "organization center" of the amphibian embryo: Its origin, spatial organization and morphogenetic action, *Advances in Morphogenetics* 10(1973):1–310.

p. 109 **fond memories of Hans Spemann:** DeRobertis, et al. (2000); E.M. Agius, O. Oelgeschläger, C. Wessely, et al., Endodermal Nodal-related signals and mesoderm induction in *Xenopus*, *Development* 127(2000):1151–1159; Gilbert, *Developmental Biology*, p. 326.

p. 109 **that will form the head structures:** Ibid., p. 313.

p. 110 **unless it's told otherwise:** A. Hemmati-Brivanlou and D.A. Melton, Vertebrate embryonic cells will become nerve cells unless told otherwise, *Cell* 88(1997):13–17.

p. 110 **synonyms . . . at the ectoderm:** DeRobertis, et al. (2000); E.L. Ferguson, Conservation of dorsal-ventral patterning in arthropods and chordates, *Current Opinion in Genetics & Development* 6(1996):424–431; S.A. Holley and E.I. Ferguson, Fish are like flies are like frogs: Conservation of dorsal-ventral patterning mechanisms, *BioEssays* 19(1997):281–284.

p. 110 **"Chordin! . . . Or "follistatin!":** DeRobertis, et al. (2000); Ferguson (1996); Holley and Ferguson (1997); W.C. Smith and R.M. Howland, Expression cloning of noggin, a new dorsalizing factor localized to the Spemann organizer in *Xenopus* embryos, *Cell* 70(1992):829–840; Y. Sasai, B. Lu, H. Steinbeisser, et al., *Xenopus* chordin: A novel dorsalizing factor activated by organizer-specific homeobox genes, *Cell* 79(1994):779–790; A. Hemmati-Brivanlou and D.A. Melton, Inhibition of activin signaling promotes neutralization in *Xenopus*, *Cell* 77(1994):273–281.

p. 110 **tissues of the placenta:** Gilbert, *Developmental Biology*, pp. 345–384.

p. 111 **the dorsal side of a chick embryo:** Ibid.

p. 111 **neural ectoderm or epidermal ectoderm:** Holley and Ferguson (1997); R.S.P. Beddington and J.C. Smith, Control of vertebrate gastrulation: Inducing signals and responding genes, *Current Opinion in Genetics & Development* 3(1993):655–661.

p. 111 **the induction of the mesoderm:** A. Streit, A.J Berliner, C. Papnayoutou, et al., Initiation of neural induction by FGF signaling before gastrulation, *Nature* 406(2000):74–78; J. Akai and K. Storey, Brain or brawn: How FGF signaling gives us both, *Cell* 115(2003):510–512.

p. 111 **"The most important event of your life":** L. Wolpert, 1986, quoted in S.F. Gilbert, *Developmental Biology*, 4th ed. (Sunderland, MA: Sinauer Associates, 1994), p. 202.

p. 113 **"Setting up a graded positional cue":** M. Osterfield, M.W. Kirschner, and J.G. Flanagan, Graded positional information: Interpretation for both fate and guidance, *Cell* 113(2003):425–428.

p. 113 **"it is not surprising that evolution":** Ibid.

p. 114 **begins life as part of a larger precursor:** P.W. Ingham and A.P. McMahon, Hedgehog signaling in animal development; Paradigms and principles, *Genes & Development* 15(2001):3059–3087; J.J. Lee, S.C. Ekker, D.P. von Kessler, et al., Autoproteolysis in hedgehog protein biogenesis, *Science* 266(1994):1528–1537.

p. 114 **and a molecule of cholesterol:** P.W. Ingham, Hedgehog signaling: A tale of two lipids, *Science* 294(2001):1879–1881; J.A. Porter, K.E. Young, and P.A. Beachy, Cholesterol modification of hedgehog signaling proteins in animal development, *Science* 274(1996):255–259; R. Nusse, Wnts and Hedgehogs: Lipid-modified proteins and similarities in signaling mechanisms at the cell surface, *Development* 130(2003):5297–5305.

p. 115 **exactly where it's needed:** Ingham and McMahon (2001); Nusse (2003); R. Burke, D. Nellen, M. Bellotto, et al., Dispatched, a novel sterol-sensing domain protein dedicated to the release of cholesterol-modified hedgehog from signaling cells, *Cell* 99(1999):803–815.

p. 115 **so is its receptor, Patched:** Ingham and McMahon (2001); Nusse (2003); J. Quirk, M. van den Huevel, and D. Henrique, The *smoothened* gene and Hedgehog signal transduction in *Drosophila* and vertebrate development, *Cold Spring Harbor Symposia* 62(1997):217–226.

p. 115 **rather than a gene activator:** Ingham and McMahon (2001); Nusse (2003).

p. 116 **known as the neural plate:** Gilbert, *Developmental Biology*, pp. 391–402.

p. 117 **grown up and done with baby talk:** R.E. Keller, Vital dye mapping of the gastrula and neurula of *Xenopus laevis*. II. Prospective areas and morphogenetic movements of the deep layer, *Developmental Biology* 51(1976):118–137.

p. 118 **pointing in a ventral-to-dorsal direction:** Y. Tanabe and T.M. Jessell, Diversity and pattern in the developing spinal cord, *Science* 274(1996):1115–1123; T.M. Jessell, Neuronal specification in the spinal cord: Inductive signals and transcriptional codes, *Nature Reviews Genetics* 1(2000):20–29.

p. 118 **the bone morphogenetic proteins BMP4 and BMP7:** Tanabe and Jessell (1996); Jessell (2000); K.F. Liem Jr., T.M. Jessel, and J. Briscoe, Regulation of the neural patterning activity of Sonic hedgehog by secreted BMP inhibitors expressed by notochord and somites, *Development* 127(2000):4855–4866.

p. 118 **"Contingency plans are part and parcel of development":** E.C. Lai, Notch signaling: Control of cell communication and cell fate, *Development* 131(2004):965–973.

p. 119 **daring to expose only its head and shoulders . . . in one gulp:** S. Artavanis-Tsakonas, M.D. Rand, and R.J. Lake, Notch signaling: Cell fate control and signal integration in development, *Science* 284(1999):770–776; G. Weinmaster, The ins and outs of Notch signaling, *Molecular and Cellular Neuroscience* 9(1997): 91–10; G. Weinmaster, Notch signaling: Direct or what? *Current Opinion in Genetics & Development* 8(1998):436–442; R. Kopan, Notch: A membrane-bound transcription factor, *Journal of Cell Science* 115(2002):1095–1097.

p. 119 **regulating so-called proneural genes:** Lai (2004); Weinmaster (1997); Kopan (2002).

p. 119 **touch receptors in our skin:** J.B. Skeath and S.B. Carroll, Regulation of proneural gene expression and cell fate during neuroblast segregation in the *Drosophila* embryo, *Development* 114(1992):39–946; A.L. Parks and M.A.T. Muskavitch, *Delta* function is required for bristle organ determination and morphogenesis in *Drosophila melanogaster*, *Developmental Biology* 157(1993):484–496; P. Heitzler and P. Simpson, The choice of cell fate in the epidermis of *Drosophila*, *Cell* 64(1991):1083–1092; I. Greenwald and G.M. Rubin, Making a difference: The role of cell-cell interactions in establishing separate identities for equivalent cells, *Cell* 68(1992):271–281.

p. 121 **which will change their minds:** T. Gridley, Notch signaling in vertebrate development and disease, *Molecular and Cellular Neuroscience* 9(1997):103–108.

p. 121 **with superfluous neurons:** A. Chitnis, D. Henrique, J. Lewis, et al., Primary neurogenesis in *Xenopus* embryos regulated by a hologue of the *Drosophila* neurogenic gene *Delta*, *Nature* 375(1995):761–766.

p. 122 **spitting out tickets at the end of each round:** O. Pourquié, The segmentation clock: Converting embryonic time into spatial pattern, *Science* 301(2003):328–330.

p. 122 **the center of the mechanism:** Pourquié (2003); C. Jouve, T. Imura, and O. Pourquié, Onset of segmentation clock in the chick embryo: Evidence for oscillations in the somite precursors in the primitive streak, *Development* 129(2002):1107–1117.

p. 122 **that suppresses Notch signaling:** Pourquié (2003).

p. 122 **the growth factor FGF:** Pourquié (2003); J. Dubrelle, M.J. McGrew, and O. Pourquié, FGF signaling controls somite boundary position and regulates segmentation clock control of spatiotemporal Hox gene activation, *Cell* 106(2001): 219–232; A. Sawada, M. Shinya, Y.-J. Jiang, et al., Fgf/MAPK signaling is a crucial positional cue in somite boundary formation, *Development* 128(2001): 4873–4880.

p. 123 **in particular the Wnt proteins:** H.M. Stern, A.M.C. Brown, and S.D. Hauschka, Myogenesis in paraxial mesoderm: preferential induction by dorsal neural tube and by cells expressing *Wnt-1*, *Development* 121(1995):3675–3686; M. Ikeya and S. Takada, Wnt signaling from the dorsal neural tube is required for the formation of the medial dermomyotome, *Development* 125(1998):4969–4976.

p. 123 **the first Wnt to be discovered:** K.M. Cadigan and R. Nusse, Wnt signaling: A common theme in animal development, *Genes & Development* 11(1997): 3286–3305. An excellent resource for more information on the Wnt signaling pathway is the Wnt Gene Homepage, online at http://www.stanford.edu/rnusse/wntwindow.html.

p. 123 **embellished with a fatty acid, palmitate:** Nusse (2003).

p. 123 **spelling out "dorsal" in the amphibian embryo:** Cadigan and Nusse (1997); W.J. Nelson and R. Nusse, Convergence of Wnt, β-catenin, and cadherin pathways, *Science* 303(2004):1483–1487; A. Wodarz and R. Nusse, Mechanisms of Wnt signaling in development, *Annual Review of Cell and Developmental Biology* 14(1998):59–88.

p. 123 **before it ever gets near the nucleus:** Cadigan and Nusse (1997); Wodarz and Nusse (1998).

p. 123 **disperses the Degradation Complex:** Ibid.

p. 123 **the transcription factors MyoD and Myf-5:** Gerhart and Kirschner, *Cells, Embryos, and Evolution,* pp. 254–260; Gilbert, *Developmental Biology*, pp. 473–474.

p. 124 **every time they say "FGF":** Martin (1998).

p. 124 **each has its own nuance:** D.M. Ornitz and N. Itoh, Fibroblast growth factors, *Genome Biology* 2(2001): reviews3005.1–3005.12, online at http://genomebiology.com/2001/2/3/reviews/3005.

p. 124 **"fibroblast growth factor eight" (FGF8):** P.H. Crossley, G. Minowada, C.A. MacArthur, et al., Roles for FGF8 in the induction, initiation, and maintenance of chick limb development, *Cell* 84(1996):127–136.

p. 124 **it has two things to say:** Ohuchi, et al. (1997).

p. 125 **just above the apical ectodermal ridge:** Martin (1998); G. Martin, Making a vertebrate limb: New players enter from the wings, *BioEssays* 23 (2001): 856–868; R.D. Riddle, M. Ensini. C. Nelson et al., Induction of the LIM homeobox gene Lmx 1 by Wnt7a establishes dorsoventral pattern in the vertebrate limb, *Cell* 83(1995):631–640.

p. 125 **where the bud leaves the body wall:** R.D. Riddle, R.L. Johnson, E. Laufer, et al., *Sonic hedgehog* mediates the polarizing activity of the ZPA, *Cell* 75(1993):1401–1416.

p. 125 **to keep the words flowing:** Vogel, et al. (1996); Ohuci, et al. (1995); Crossley, et al. (1996); L. Niswander, S. Jeffrey, G.R. Martin, et al., A positive feedback loop coordinates growth and patterning in the vertebrate limb, *Nature* 371(1994):609–612.

p. 125 **the proximal-distal pattern of the developing limb:** D. Summerbell, J.H. Lewis, and L. Wolpert, Positional information in chick limb morphogenesis, *Nature* 244(1973):492–496.

p. 127 **stopped them in their tracks:** A.T. Dudley, M.A. Ros, and C.J. Tabin, A re-examination of proximodistal patterning during vertebrate limb development, *Nature* 418(2002):539–544.

p. 129 **a death that was quiet, orderly, even dignified:** J.F.R. Kerr, A.H. Wylie, and A.R. Currie, Apoptosis: A basic biological phenomenon with wide ranging implications in tissue kinetics, *British Journal of Cancer* 26(1972):239–257.

p. 129 **the availability of precious resources:** R.E. Ellis, J. Yuan, and H.R. Horvitz, Mechanisms and function of cell death, *Annual Review of Cell Biology* 7(1997):663–698.

p. 130 **scavenge the remains of the dead one:** M. Barinaga, Death by dozens of cuts, *Science* 280(1998):32–34; N.A. Thornberry and Y. Lazebnik, Caspases: Enemies within, *Science* 28(1998):1312–1316; M.O. Hengartner, The biochemistry of apoptosis, *Nature* 407(2000):770–776.

p. 130 **"Caspases are members of the society too":** Y. Lazebnik, "Enemies Within or the Biochemistry of Apoptosis," presentation at the 164th national meeting of the American Association for the Advancement of Science, February 16, 1998.

p. 130 **the demise of the enzyme's unfortunate landlord:** Hengartner (2000); D.D. Newmeyer and S. Ferguson-Miller, Mitochondria: Releasing power for life and unleashing the machineries of death, *Cell* 112(2003):481–490; D.R. Green and J.C. Reed, Mitochondria and apoptosis, *Science* 281(1998):1309–1311.

p. 131 **cell savers Bcl-2 and Bcl-x_L:** Hengartner (2000); Newmeyer and Ferguson-Miller (2003); T. Chittenden, BH3 domains: Intracellular death-ligands critical for initiating apoptosis, *Cancer Cell* 2(2002):165–166; J.M. Adams and S. Cory, The Bcl-2 protein family; Arbiters of cell survival, *Science* 281(1998):1322–1325.

p. 131 **a cell's decision to live or die:** Ellis, et al. (1997); G. Evan and T. Littlewood, A matter of life and cell death, *Science* 281(1998):1317–1321.

p. 131 **the collective suicide of thousands of cells:** L. Grotewold and U. Ruther, Bmp, Fgf, and Wnt signaling in programmed cell death and chondrogenesis during vertebrate limb development: The role of Dickkopf-1, *International Journal of Developmental Biology* 46(2002):943–947; R. Merino, Y Gañán, D. Macias, et al., Bone morphogenetic proteins regulate interdigital cell death in the avian embryo, *Annals of the New York Academy of Sciences* 887(1999):120–132; Y. Yokouchi, J. Sakiyama, T. Kameda, et al., BMP-2/-4 mediate programmed cell death in chicken limb buds, *Development* 122(1996):3725–3724; H. Zou and L. Niswander, Requirement for BMP signaling in interdigital apoptosis and scale formation, *Science* 272(1996):738–741.

p. 131 **A ring of death . . . behind a BMP antagonist:** Merino, et al. (1999); Yokouchi, et al. (1996); Zou and Niswander (1996); Y. Gañán, D. Macias, M. Duterque-Coquillaud, et al., Role of TGFP and BMPs as signals controlling the position of the digits and the areas of cell death in the developing chick autopod, *Development* 122(1996):2349–2357; R. Merino, Y. Gañán, D. Macias, et al., Morphogenesis of digits in the avian limb is controlled by FGFs, TGFβs, and Noggin through BMP signaling, *Developmental Biology* 200(1998):35–45.

p. 131 **the cartilage template for future bone:** Merino, et al. (1999); Merino, et al. (1998).

p. 132 **BMP actually means "build":** Grotewold and Ruther (2002); Merino, et al. (1998).

p. 132 **"BMP" means "go kill yourself":** Grotewold and Ruther (2002); Merino, et al. (1999); J.A. Montero, Y. Gañán, D. Macias, et al., Role of FGFs in the control of programmed cell death during limb development, *Development* 128(2001):2075–2084.

p. 132 **"is a social phenomenon":** M.C. Raff, "Death wish," *The Sciences* July/August (1996):36–40.

p. 132 **as well as the MAP kinases:** L.C. Cantley, The phosphoinositide 3-kinase pathway, *Science* 296(2002):1655–1657.

p. 133 **half of which are actually needed:** M.C. Raff, B.A. Barres, J.F. Burne, et al., Programmed cell death and the control of cell survival: Lessons from the nervous system, *Science* 262(1993):695–700.

p. 133 **out of the head like rising bread dough:** K. Kuida, et al., Decreased apoptosis in the brain and premature lethality in CPP32-deficient mice, *Nature* 384(1996):368–372.

p. 133 **face certain death by apoptosis:** J. Yuan and B.A. Yankner, Apoptosis in the nervous system, *Nature* 407(2000):802–809.

p. 133 **Discovered in the 1950s by Rita Levi-Montalcini:** S. Cohen, R. Levi-Montalcini, and V. Hamburger, A nerve growth-stimulating factor isolated from sarcomas 37 and 180, *Proceedings of the National Academy of Sciences USA* 40(1954):1014–1018.

p. 133 **binds to a receptor tyrosine kinase called TrkA:** E. Huang and L. Reichardt, Trk receptors: Roles in neuronal signal transduction, *Annual Review of Biochemistry* 72(2003):609–642.

p. 134 **Instead, the TrkA receptor . . . to act as the courier:** F.D. Miller and D.R. Kaplan, TRK makes the retrograde, *Science* 295(2002):1471–1473; B.L. MacInnis and R.B. Campenot, Retrograde support of neuronal survival without retrograde transport of nerve growth factor, *Science* 295(2002):536–1539; D.L. Senger and R.B. Campenot, Rapid retrograde tyrosine phosphorylation of trkA and other proteins in rat sympathetic neurons in compartmented cultures, *Journal of Cell Biology* 138(1997):411–421.

p. 134 **and kill themselves in despair:** Dudley, et al. (2002).

p. 136 **"The embryo inherits a rather compact 'tool kit'":** Gilbert, *Developmental Biology*, pp. 149–150.

p. 136 **"Genes expressed during formation of one part":** M. Ptashne and A. Gann, *Genes and Signals* (Cold Spring Harbor, NY: Cold Spring Harbor Press, 2002), p. 3.

p. 136 **even this lowly creature . . . can mouth a Wnt sentence:** B. Hobmayer,

F. Rentzsch, K. Kuhn, et al., WNT signaling molecules act in axis formation in the diploblastic metazoan hydra, *Nature* 407(2000):186–189.

p. 136 **the Wnt signaling relay, in plants:** V. Amador, E. Monte, J. Garcia-Marinez, et al., Gibberellins signal nuclear import of PHOR1, a photoperiod-responsive protein with homology to *Drosophila* armadillo, *Cell* 106(2001):343–354.

p. 137 **fly wings by Hedgehog and Wingless:** M. Stringini and S.M. Cohen, Formation of morphogen gradients in the *Drosophila* wing, *Seminars in Cell and Developmental Biology* 10(1999):335–344.

p. 137 **30,000 different words to compose his plays:** "A man of many words," online at http://shakespeare.about.com/b/a/020320.html.

4
LIFE IN THE BALANCE

p. 140 **"the fell blows of circumstance":** W. Cannon, *The Wisdom of the Body* (New York: W.W. Norton, 1929), p. 244.

p. 140 **"continue to live and carry on their functions":** Cannon, *Wisdom*, pp. 22–23.

p. 140 **"the coordinated physiological processes":** Cannon, *Wisdom*, p. 24.

p. 140 **possible only because biological processes are so elastic:** Cannon, *Wisdom*, pp. 21, 24.

p. 141 **the exchange of chemical signals was also apparent:** Cannon, *Wisdom*, p. 255.

p. 141 **"what influences the signal" . . . "has disclosed the facts":** Cannon, *Wisdom*, p. 289.

p. 141 **"the unstable stuff of which we are composed":** Cannon, *Wisdom*, p. 23.

p. 142 **at about 120 to be exact:** B. Alberts, A. Johnson, J. Lewis, M. Raff, K. Roberts, and P. Walter, *Molecular Biology of the Cell*, 4th ed. (New York: Garland Science), p. 1292.

p. 143 **are not always identical twins:** Ibid., p. 1262; C.S. Potten, M. Loeffler, Stem cells: Attributes, cycles, spirals, pitfalls, and uncertainties. Lessons for and from the crypt, *Development* 110(1990):1001–1020.

p. 144 **"than any other tissue in the body":** Alberts, et al., *Molecular Biology of the Cell*, p. 1260.

p. 144 **"You there!" they shout . . . "Platelet-derived growth factor!":** T.D. Pollard and W.C. Earnshaw, *Cell Biology* (Philadelphia: Saunders, 2002), pp. 547–548; S. Werner and R. Grose, Regulation of wound healing by growth factors and cytokines, *Physiological Reviews* 83(2003):835–870; W.J.H. Kim, Cellular signaling during tissue regeneration, *Yonsei Medical Journal* 41(2000):692–703; R.A. Clark, Regulation of fibroplasia in cutaneous wound repair, *American Journal of Medical Science* 306 (1993):42–48.

p. 144 **fibroblast growth factors FGF2 and FGF7:** Werner and Grose (2003); Kim (2000).

p. 144 **pull the edges of the wound together:** Pollard and Earnshaw, *Cell Biology*, pp. 547–548; Werner and Grose (2003); Clark (1993).

p. 144 **the keratinocytes and their progenitors:** Werner and Grose (2003).

p. 144 **encouraging them to divide:** Werner and Grose (2003); Kim (2000).

p. 144 **raise a roof over the wound:** Werner and Grose (2003); Kim (2000); G. Zambruno, P.C. Marchisio, A. Marconi, et al., Transforming growth factor-beta 1 modulates beta 1 and beta 5 integrin receptors and induces the de novo expression of the alpha v beta 6 heterodimer in normal human keratinocytes: Implications for wound healing, *Journal of Cell Biology* 129(1995):853–865.

p. 145 **an entire network of new capillaries:** Werner and Grose (2003); S. Frank, G. Hübner, G. Breier, et al., Regulation of vascular endothelial growth factor expression in cultured keratinocytes: Implications for normal and impaired wound healing, *Journal of Biological Chemistry* 270(1995):12607–12613.

p. 145 **poked around the primordial aorta:** S.F. Gilbert, *Developmental Biology*, 7th ed. (Sunderland, MA: 2003), p. 507; M. Kondo, A.J. Wagers, and M.G. Manz, Biology of hematopoietic stem cells and progenitors: Implications for clinical application, *Annual Review of Immunology* 21(2003):759–806.

p. 146 **give up their wandering ways:** Kondo, et al. (2003); D.E. Wright, A.J. Wagers, and A.P. Gulati, Physiological migration of hematopoietic stem and progenitor cells, *Science* 294(2001):1933–1936.

p. 146 **dividing and replenishing:** T. Reya, A.W. Duncan, L. Ailles, et al., A role for Wnt signaling in self-renewal of haematopoietic stem cells, *Nature* 423(2003): 409–414.

p. 146 **another synonym for "stay young":** F.N. Karanu, B. Murdoch, L. Gallacher, et al., The notch ligand jagged-1 represents a novel growth factor of human hematopoietic stem cells, *Journal of Experimental Medicine* 192(2000):1365–1372; B. Varnum-Finney, L. Wu, M. Yu, et al., Pluripotent, cytokine-dependent, hematopoietic stem cells are immortalized by constitutive Notch1 signaling, *Nature Medicine* 6(2000):1278–1281.

p. 146 **Sonic hedgehog is a third:** G. Bhardwaj, B. Murdoch, D. Wu, et al., Sonic hedgehog induces the proliferation of primitive human hematopoietic cells via BMP regulation, *Nature Immunology* 2(2001):172–180.

p. 146 **as much as tenfold:** Kondo, et al. (2003).

p. 147 **of that fateful choice:** Kondo, et al. (2003); A.J. Wagers, J.L. Christensen, and I.L. Weissman, Cell fate determination from stem cells, *Gene Therapy* 9(2002):606–612; Gilbert, *Developmental Biology*, pp. 508–509; Alberts, et al., *Molecular Biology of the Cell*, pp. 1283–1296.

pp. 147–148 **rather than endocrine glands:** W.J. Leonard, Type I cytokines and interferons and their receptors, in W.E. Paul (ed), *Fundamental Immunology*, 4th ed. (Philadelphia: Lippencott-Raven, 1999), pp. 741–774.

p. 148 **is a separate protein:** D.E. Levy and J.E. Darnell, Jr., STATs: Transcriptional control and biological impact, *Nature Reviews Cell and Molecular Biology* 3(2002):651–662; A. Ziemiecki, A.G. Harpur, and A.F. Wilks, Jak protein tyrosine kinases: Their role in cytokine signaling, *Trends in Cell Biology* 4(1994):207–212; C. Schinder and J.E. Darnell, Jr., Transcriptional responses to polypeptide ligands: the Jak-STAT pathway, *Annual Review of Biochemistry* 64(1995):621–651; J.N. Ihle, B.A.

Witthuhn, F.W. Quell, et al., Signaling through the hematopoietic cytokine receptors, *Annual Review of Immunology* 13(1995):369–398.

p. 148 **interleukin-3, and interleukin-9:** Gilbert, *Developmental Biology*, p. 508.

p. 148 **secreted by the kidney:** E. Goldwasser, Erythropoietin: A somewhat personal history, *Perspectives in Biology and Medicine* 40(1996):18–32; J.W. Fisher, A quest for erythropoietin over nine decades, *Annual Review of Pharmacology and Toxicology* 38(1998):1–20; W.S. Alexander, Cytokines in hematopoiesis, *International Review of Immunology* 16(1998):651–682.

p. 149 **quickly lose their will to live:** Reya, et al. (2003).

p. 149 **role in stem cell apoptosis:** Wagers, et al. (2002).

p. 150 **stem cells of normal mice:** J. Domen and I.L. Weissman, Hematopoietic stem cells need two signals to prevent apoptosis; BCL-2 can provide one of these, Kitl/c-Kit signaling the other, *Journal of Experimental Medicine* 192(2000):1707–1718; J. Domen, S.H. Cheshier, and I.L. Weissman, The role of apoptosis in the regulation of hematopoietic stem cells: Overexpression of Bcl-2 increases both their number and repopulation potential, *Journal of Experimental Medicine* 191(2000):253–264.

p. 150 **"the body has no protection":** Cannon, *Wisdom*, p. 319.

p. 150 **veterinarians call it:** Interview transcript, John Incardona, in *Rediscovering Biology*: Unit 7, Genetics of Development, online at www.learner.org/channel/courses/biology/units/gendev/experts/incardona.html.

p. 151 **incidence of cyclopia:** Ibid.

p. 151 **appropriately enough, "cyclopamine":** Ibid.; R.F. Keeler and W. Binns, Teratogenic compounds of *Veratrum californicum* (Durand). V. Comparison of cyclopean effects of steroidal alkaloids from the plant and structurally related compounds from other sources, *Teratology* 1(1968):5–10.

p. 151 **also born with cyclopia:** C. Chiang, Y. Litingtung, E. Lee, et al., Cyclopia and defective axial patterning in mice lacking *Sonic hedgehog* gene function, *Nature* 383(1996):407–413.

p. 151 **the mirror image of cyclopia:** "Basal cell nevus syndrome," Diseases and Conditions, online at http://health.allrefer.com/health/basal-cell-nevus-syndrome-info.html.

p. 151 **signaling pathway is never quiet:** A.E. Bale and K.P. Yu, The hedgehog pathway and basal cell carcinoma, *Human Molecular Genetics* 10(2001):757–762; H. Hahn, et al., Mutations of the human homolog of *Drosophila patched* in the nevoid basal cell carcinoma syndrome, *Cell* 85(1996):841–851; R.L. Johnson, et al., Human homolog of *patched*, a candidate gene for the basal cell nevus syndrome, *Science* 272(1996):1668–1671.

p. 152 **"and its somatic milieu":** D.R. Green and G.E. Evan, A matter of life and death, *Cancer Cell* 1(2002):19–30.

p. 153 **in two ways:** Alberts, et al., *Molecular Biology of the Cell*, p. 1333; D. Hanahan and R.A. Weinberg, The hallmarks of cancer, *Cell* 100(2000):57–70; B.A.J. Ponder, Cancer genetics, *Nature* 411(2001):336–341.

p. 153 **legitimate growth signals:** P. Blume-Jensen and T. Hunter, Oncogenic kinase signaling, *Nature* 411(2000):355–365.

p. 153 **one out of every four human cancers:** Alberts, et al., *Molecular Biology of the Cell*, p. 1335; M. Barbacid, Ras genes, *Annual Review of Biochemistry* 56(1983):779–827.

p. 153 **and multiple myeloma:** Blume-Jensen and Hunter (2000).

p. 153 **no longer begins with Hedgehog:** Aberrations in Hedgehog signaling have now been implicated in a number of other cancers as well, including cancers of the digestive tract, prostate, lung, and bladder. An overview of the role of the Hedgehog pathway in cancer can found in Beachy, Karhadkar, and Berman (2004).

p. 154 **"They are the only cells . . . that live long enough":** Cancer as a stem cell disease. For a recent review of the "cancer stem cell" concept, see P.A. Beachy, S.S. Karhadkar, and D.M. Berman, Tissue repair and stem cell renewal in carcinogenesis, *Nature* 432(2004):324–331.

p. 154 **The fourth most common type of cancer:** "Cancer Facts and Figures 2004," American Cancer Society, online at www.cancer.org.

p. 154: **mitogen and morphogen Wnt:** B. Vogelstein and K.W. Kinzler, The multistep nature of cancer, *Trends in Genetics* 9(1993):138–141; K.W. Kinzler and B. Vogelstein, Lessons from hereditary colorectal cancer, *Cell* 87(1996):159–170.

p. 154 **70 percent of colorectal cancers:** Alberts, et al., *Molecular Biology of the Cell*, p. 1352.

p. 155 **$2,200 to repair the damage:** "White-tailed Deer in Pennsylvania," report by the Bureau of Wildlife management, Pennsylvania Game Commission, July 2003, online at http://www.pcg.state.pa.us/lib/pgc/deer/pdf/Management_Plan6.03.pdf.

p. 155 **appease frustrated sportsmen:** Ibid.

p. 155 **Today, 1.5 million deer:** Ibid.

p. 155 **into a space big enough for thirty:** "Deer Info Page," The Pennsylvania Audubon Society, online at http://pa.audobon.org/deerinfopage.html.

p. 155 **have more babies:** "Deer Biology," presentation by Dr. Gary Alt at the Conference on the Impact of Deer on the Biodiversity and Economy of the State of Pennsylvania, January 6, 2004, transcript available online at http://pa.audobon.org/chapter/pa/pa/Alt.html.

p. 156 **twins or even triplets every spring:** Pennsylvania Game Commission, "White-tailed Deer" (2003).

p. 156 **"The cancer cell is a renegade":** R. Weinberg, *One Renegade Cell* (New York: Basic Books, 1999), p. 97.

p. 157 **command them to commit suicide:** Green and Evan (2002); G.I. Evan and K.H. Vousden, Proliferation, cell cycle and apoptosis in cancer, *Nature* 411(2001):342–348.

p. 157 **blocking the death program:** Ibid.

p. 158 **on which all cancers are built:** Green and Evan (2002).

p. 158 **"If it were not permitted to expose our lives":** G. Minois, *History of Suicide: Voluntary Death in Western Culture*, translated by L.G. Cochrane (Baltimore: The Johns Hopkins University Press, 1999), p. 71.

p. 160 **protect the community at large:** B. Vogelstein, D. Lane, and A.J. Levine, Surfing the p53 network, *Nature* 408(2000):307–310; A.J. Levine, p53, the cellular gatekeeper for growth and division, *Cell* 88(1997):323–331.

p. 160 **almost as soon as it's off the ribosome:** Vogelstein, et al. (2000).

p. 160 **of its civic duty to die:** Vogelstein, et al. (2000); N.E. Sharpless and R.A. DePinho, p53: Good cop/bad cop, *Cell* 110(2002):9–12; J.C. Reed, Dysregulation of apoptosis in cancer, *Journal of Clinical Oncology* 17(1999):2941–2953.

p. 160 **kill themselves the next:** Green and Evan (2002); Evan and Vousden (2001); G. Evan and T. Littlewood, A matter of life and cell death, *Science* 281(1998): 1317–1322.

p. 160 **PIP3 as its direct object:** Green and Evan (2002); L.C. Cantley, The phosphoinositide 3-kinase pathway, *Science* 296(2002):1655–1657; also available online at *Science*'s Signal Transduction Knowledge Environment website, http://www.stke.org/cgi/cm.CMP_6557; A. Kauffmann-Zeh, P. Rodriguez-Viciana, E. Ulrich, et al., Suppression of c-Myc-induced apoptosis by Ras signaling through PI(3)K and PKB, *Nature* 385(1997):544–548.

p. 160 **operating the cell cycle:** D.L. Levens, Reconstructing MYC, *Genes & Development* 17(2003):1071–1077.

p. 160 **dividing cells to the police:** Green and Evan (2002); Evan and Vousden (2001); Evan and Littlewood (1998).

p. 160 **certain to trigger suicidal ideation:** Ibid.

p. 160 **releasing deadly cytochrome c:** P. Juin, A.O. Hueber, T. Littlewood, and G. Evan, c-Myc-induced sensitization to apoptosis is mediated through cytochrome c release, *Genes & Development* 13(1999):1367–1381.

p. 160 **they cut so beautifully:** Green and Evan (2002); Evan and Vousden (2001); Evan and Littlewood (1998).

p. 162 **researchers call "death ligands":** A. Ashkenazi and V.M. Dixit, Death receptors: Signaling and modulation, *Science* 281(1998):1305–1308.

p. 163 **resembled the worm protein:** N.A. Thornberry and Y. Lazebnik, Caspases: Enemies within, *Science* 281(1998):1312–1316; M. Barinaga, Forging a path to cell death, *Science* 273(1996):735–737.

p. 163 **discovery of the first caspase:** Barinaga (1996).

p. 163 **molecular arms dealers:** Ashkenazi and Dixit (1998).

p. 163 **cell survival rather than cell death:** B.C. Barnhart and M.E. Peter, The TNF receptor 1: A split personality complex, *Cell* 114(2003):148–150. O. Micheau and J. Tschopp, induction of TNF receptor 1-mediated apoptosis via two sequential signaling complexes, *Cell* 114(2003):181–190.

p. 164 **a second signaling complex:** Ibid.

p. 164 **compromising treatment:** R.W. Johnstone, A.A. Ruefli, and S.W. Lowe, Apoptosis; A link between cancer genetics and chemotherapy, *Cell* 108(2002):153–164; A.S. Baldwin, Control of oncogenesis and cancer therapy resistance by the transcription factor NF-κB, *Journal of Clinical Investigation* 107(2001):241–246.

p. 164 **shielding the tumors from its lethal effects:** R.M. Pitti, S.A. Marsters, D.A. Lawrence, et al., Genomic amplification of a decoy receptor for Fas ligand in lung and colon cancer, *Nature* 396(1998):699–703. Mutations in death ligand signaling pathways are not the only way to derail apoptosis, freeing a cell to pursue delusions of immortality. Overexpression of anti-apoptotic factors like Bcl-2, for example, upsets the balance of power at the mitochondrial membrane; as a result, even a cell that's well aware that it's on the road to malignancy cannot access cytochrome c. The name "Bcl," in fact, comes from the discovery of such mutations in human follicular *B c*ell *l*ymphoma, a blood cancer. See Green and Evan (2002) and Y. Tsujimoto, L.R. Finger, J. Yunis, et al., Cloning of the chromosome breakpoint of neoplastic B cells with the t(14; 18) chromosome translocation, *Science* 226(1984):1097–1099.

p. 167 **"Insofar as the constancy of the fluid matrix":** Cannon, *Wisdom*, p. 302.

p. 167 **can also poison them:** M. Brownlee, Biochemistry and molecular cell biology of diabetic complications, *Nature* 414(2001):813–820.

p. 167 **retina, kidney, and peripheral nerves:** Ibid.

p. 168 **glycogen or fatty acids:** J.E. Pessin and A.R. Saltiel, Signaling pathways in insulin action: Molecular targets of insulin resistance, *Journal of Clinical Investigation* 106(2000):165–169; J.E.-B. Reusch, Focus on insulin resistance in type 2 diabetes: Therapeutic implications, *Diabetes Educator* 24(1999):188–193; B.B. Kahn and J.S. Flier, Obesity and insulin resistance, *Journal of Clinical Investigation* 106(2000):473–481.

p. 168 **18 million Americans:** American Diabetes Association, "National Diabetes Fact Sheet," online at http://www.diabetes.org/diabetes-statistics/national-diabetes-fact-sheet-jsp.

p. 168 **metabolic disease in the world:** A. Dove, Seeking sweet relief for diabetes, *Nature Biotechnology* 20(2002):977–981.

p. 168 **increases the risk of heart attack . . . limb amputation:** Diabetes Fact sheet; Dove (2002).

p. 168 **an estimated $100 billion annually:** Diabetes Fact Sheet.

p. 168 **children . . . account for an increasing proportion:** Diabetes Fact Sheet; American Diabetes Association, "Diabetes Facts and Figures," online at http://www.diabetes.org/ada/facts.asp.

p. 168 **expected to double over the next 25 years:** Dove (2002).

p. 169 **discovered the pancreas:** "History of the Pancreas," online at http://www.pancreasclub.com/history.

p. 169 **in the fourth century BC:** Ibid.

p. 169 **date back to 1500 BC:** Ibid.; Canadian Diabetes Association, "The History of Diabetes," online at www.diabetes.ca/Section_about/timeline.asp.

p. 169 **coursing through the trunk:** "History of the Pancreas"; A.Q. Maisel, *The Hormone Quest* (New York: Random House, 1965), p. 6.

p. 169 **initial segment of the small intestine:** "History of the Pancreas."

p. 170 **was an exocrine gland:** Ibid.

p. 170 **under the microscope:** Ibid.; M. Bliss, *The Discovery of Insulin* (Chicago: The University of Chicago Press, 1982), p. 25.

p. 170 **the unfortunate animal developed diabetes:** Maisel, *The Hormone Quest*, p. 24; Bliss, *Discovery of Insulin*, pp. 25–26; C. Singer and E.A. Underwood, *A Short History of Medicine* (New York: Oxford University Press, 1962), p. 553.

p. 170 **the deficiency theory was correct:** Maisel, *The Hormone Quest*, pp. 25–26; Bliss, *Discovery of Insulin*, pp. 25–26; J.C. Krantz, Jr., Frederick G. Banting, Charles H. Best, and insulin, in *Historical Medical Classics Involving New Drugs* (Baltimore: Williams & Wilkins, 1974), pp. 58–65.

p. 170 **did not respond to insulin:** H.P. Himsworth, Diabetes mellitus: Its differentiation into insulin-sensitive and insulin-insensitive types, *Lancet* 1(1936):117–121.

p. 171 **rather than alarmingly low:** R.S. Yalow and S.A. Berson, Plasma insulin concentrations in nondiabeteic and early diabetic subjects, *Diabetes* 4(1960):254–260.

p. 171 **does such a great cover-up job:** A.R. Saltiel, The molecular and physiological basis of insulin resistance: Emerging implications for metabolic and cardiovascular diseases, *Journal of Clinical Investigation* 106(2000):163–164; K.S. Polonsky, J. Sturis, and G.I. Bell, Non-insulin-dependent diabetes mellitus—A genetically programmed failure of the beta cell to compensate for insulin resistance, *New England Journal of Medicine* 334(1996):777–783; B.B. Kahn, Type 2 diabetes: When insulin secretion fails to compensate for insulin resistance, *Cell* 92(1998):593–596; M.K. Cavaghan, D.A. Ehrmann, and K.S. Polonsky, Interactions between insulin resistance and insulin secretion, *Journal of Clinical Investigation* 106(2000):329–333.

p. 172 **a cartoon . . . to accompany a review article:** C.R. Kahn, What is the molecular basis for the action of insulin? *Trends in Biochemical Sciences* 4(1979): N263–N266.

p. 172 **a pair of β subunits that constitute the kinase:** A.R. Saltiel and C.R. Kahn, Insulin signaling and the regulation of glucose and lipid metabolism, *Nature* 414(2001):799–805.

p. 172 **a phosphate or two themselves:** M.F. White, The IRS-signaling system: A network of docking proteins that mediate insulin action, *Molecular and Cellular Biochemistry* 182(1998):3–11.

p. 172 **and then a second, IRS-2:** Ibid.; M.F. White, R. Maron, and C.R. Kahn, Insulin rapidly stimulates tyrosine phosphorylation of Mr-185,000 protein in intact cells, *Nature* 318(1985):183–186.

p. 173 **in the signaling relay:** M.F. White, Insulin signaling in health and disease, *Science* 302(2003):1710–1711. For an illustration of the current state of insulin signaling, see the "Insulin Signaling Pathway" Connections Map at the Science STKE website: www.stke.sciencemag.org/cgi/cm/cmp_12069.

p. 174 **fleet of membrane-bound vesicles:** P.R. Shepherd and B.B. Kahn, Glucose transporters and insulin action, *New England Journal of Medicine* 341(1999): 248–257. The GLUT4 transporter is actually one member of a group of five homologous glucose transport proteins; of the five, it is the one most important to insulin-mediated glucose uptake.

p. 174 **actin fibers of the cytoskeleton:** Saltiel and Kahn (2001).

p. 175 **IRS protein—PI3kinase—PIP3—Akt:** Ibid.; A.R. Saltiel and J.E. Pessin, Insulin signaling pathways in time and space, *Trends in Cell Biology* 12(2002):65–71; C.A. Baumann and A.R. Saltiel, Spatial compartmentalization of signal transduction in insulin action, *BioEssays* 23(2001):215–222.

p 175 **biologists call "lipid rafts":** Saltiel and Kahn (2001); Saltiel and Pessin (2002); C.A. Baumann, V. Ribon, M. Kanzaki, et al., CAP defines a second signaling pathway required for insulin-stimulated glucose transport, *Nature* 407(2000):202–207.

p. 175 **a protein called CAP:** Saltiel and Kahn (2001); Saltiel and Pessin (2002); Baumann and Saltiel (2001); Baumann, et al. (2000).

p. 175 **GLUT4 and its vesicles:** Saltiel and Kahn (2001); Shepherd and Kahn (1999); J. Alper, New insights into type 2 diabetes, *Science* 289(2000):37–39; G.I. Shulman, Cellular mechanisms of insulin resistance, *Journal of Clinical Investigation* 106(2000):171–176.

p. 176 **More than 80 percent . . . are obese:** J.S. Flier, The missing link with obesity?, *Nature* 409(2001):292–293.

p. 176 **currently thought to be overweight:** National Institute for Diabetes, Digestive and Kidney Disorders, "Prevalence statistics related to overweight and obesity," online at http://win.niddk.nih.gove/statistics/index.htm#preval. Figures quoted are for 2000. What's worse, this battle of the bulge now seems to encompass even the youngest members of society—recent data from the American Heart Association indicate that approximately 10% of children between the ages of 2 and 5 can be classified as overweight.

p. 176 **and eventually diabetes:** P. Björntorp, Abdominal obesity and the development of noninsulin-dependent diabetes mellitus, *Diabetes/Metabolism Reviews* 4(1988):615–622; H.E. Lebovitz, Pathogenesis of type 2 diabetes, *Drug Benefit Trends* 12(suppl A, 2000):8–16; S.E. Kahn, R.L. Prigeon, R.S. Schwartz, et al., Obesity, body fat distribution, insulin sensitivity and islet β-cell function as explanations for metabolic diversity, *Journal of Nutrition* 131(2001):354S–360S.

p. 176 **their response to insulin improves:** J.F. Bak, N. Moller, O. Schmitz, et al., In vivo insulin action and muscle glycogen synthase activity in type 2 (non-insulin-dependent) diabetes mellitus: Effects of diet treatment, *Diabetologia* 35(1992): 777–784.

p. 176 **blindness and kidney failure decrease:** According to the National Institute for Diabetes and Digestive and Kidney Diseases, every 1 percent reduction in the protein HbA1c (now thought to be the best indicator of glucose tolerance) results in a 40% decrease in the risk of complications. What's more, weight loss can actually delay or even prevent the development of type 2 diabetes in susceptible individuals. In a recent study of "pre-diabetic" individuals (i.e., patients with insulin resistance and impaired glucose tolerance that had not yet reached pathological levels), a loss of just 5 to 10 pounds, coupled with regular exercise, reduced the incidence of type 2 diabetes by 58% over a three-year period. See Knowler WC, E. Barrett-Connor, S.E.Fowler, et al., Reduction in the incidence of type 2 diabetes with lifestyle intervention or metformin, *New England Journal of Medicine* 346(2002):393–403.

p. 177 **$30 billion to $50 billion spent each year:** Weight-control Information Network, "Statistics related to overweight and obesity," online at http://win.niddk.nih.gov/statistics/#econ.

p. 177 **90 percent will eventually gain it all back:** J.M Friedman, A war on obesity, not the obese, *Science* 299(2003):856–858; G. Kolata, "How the Body Knows When to Gain or Lose," *New York Times*, October 17, 2000, Science Times section.

p. 178 **a prodigious appetite to go with it:** D.L. Coleman, Obese and diabetes: Two mutant genes causing diabetes-obesity syndromes in mice, *Diabetologia* 14(1978):141–148; G.A. Bray and D.A. York, Hypothalamic and genetic obesity in experimental animals: An autonomic and endocrine hypothesis, *Physiological Reviews* 59(1979):719–809.

p. 178 **an *ob/ob* mouse transfused with blood:** D.L. Coleman, Effects of parabiosis of obese with diabetes and normal mice, *Diabetologia* 9(1973):294–298.

p. 178 **succeeded in cloning the *ob* gene:** Y. Zhang, R. Proenca, M. Maffei, et al., Positional cloning of the mouse obese gene and its human homologue, *Nature* 372(1994):425–432.

p. 178 **the maturation of hematopoietic stem cells:** J.M. Friedman and J.L. Halaas, Leptin and the regulation of body weight in mammals, *Nature* 395(1998): 763–770; B.C. Moon and J.M. Friedman, The molecular basis of the obese mutation in ob^{2J} mice, *Genomics* 42(1997):152–156; T. Madej, M.S. Boguski, and S.H. Bryant, Threading analysis suggests that the obese gene product may be a helical cytokine, *FEBS Letters* 373(1995):13–18.

p. 179 **other fat cell hormones:** G. Frühbeck, J. Gómez-Ambrosi, F.J. Muruzábal, et al., The adipocyte: A model for integration of endocrine and metabolic signaling in energy metabolism, *American Journal of Physiology, Endocrinology, and Metabolism* 280(2001):E827–E847; A.R. Saltiel, You are what you secrete, *Nature Medicine* 7(2001):887–888.

p. 179 **One, adiponectin, increases sensitivity:** Saltiel (2001); P.E. Scherer, S. Williams, M. Fogliano, et al., A novel serum protein similar to C1q, produced exclusively in adipocytes, *Journal of Biological Chemistry* 270(1995):26746–26749; E. Hu, P. Liang, and B.M. Spiegelman, AdipoQ is a novel adipose-specific gene dysregulated in obesity, *Journal of Biological Chemistry* 271(1996):10697–10703; T. Yamauchi, et al., The fat-derived hormone adiponectin reverses insulin resistance associated with both lipoatrophy and obesity, *Nature Medicine* 7(2001):941–946.

p. 179 **resistin . . . triggers insulin resistance:** C.M. Steppan, S.T. Bailey, S. Bhat, et al., The hormone resistin links obesity to diabetes, *Nature* 409(2001):307–312; R.R. Banerjee, S.M. Rangwala, J.S. Shapiro, et al., Regulation of fasted blood glucose by resistin, *Science* 303(2004):1195–1198.

p. 179 **like adiponectin and resistin:** Saltiel (2001); Banerjee (2004).

p. 179 **Studies of . . . animals and . . . patients:** Saltiel (2001); Steppan (2001).

pp. 179–180 **The balancing act upset:** The discovery of resistin came about as part of Lazar's search for the mechanism of action of a novel class of oral medications for the treatment of type 2 diabetes. Known as thiazolidinediones (TZDs), or glitazones, these agents increase insulin sensitivity and bind to receptors on fat cells called PPAR-γ, for "peroxisome proliferator activated receptor, γ subvariety."

If the insulin receptor and its baroque intracellular signaling mechanisms are an example of the eloquence of mammalian cells, the PPAR-γ receptor is a testament to their brevity. Like the LuxR receptor of light-producing bacteria, PPAR-γ combines binding site and transcription factor in a single protein, a signaling strategy also employed by steroid hormones such as estrogen and progesterone. TZDs, Lazar found, bind to this so-called nuclear receptor and downregulate the expression of the resistin gene; as resistin levels fall, insulin sensitivity improves.

Unlike other anti-diabetic medications currently on the market, which block the absorption or synthesis of glucose or goad an already overworked pancreas to secrete even more insulin, TZDs make existing insulin more effective, introducing the voice of reason to addled discussions surrounding the fate of dietary glucose. "If you have a broken phone connection, you can shout louder or fix the problem. Other drugs shout louder. TZDs fix the connection," says Lazar. Unfortunately, Rezulin, the first TZD introduced in the early 1990s, had to be withdrawn by the manufacturer after reports of severe liver toxicity. However, two other TZDs with a better safety record, Avandia (GlaxoSmithKline) and Actos (Eli Lilly), are currently available and others are in the pipeline.

p. 180 **svelte in no time:** Friedman and Halaas (1998); M.A. Pellymounter, M.J. Cullen, M.B. Baker, et al., Effects of the *obese* gene product on body weight regulation in *ob/ob* mice, *Science* 269(1995):540–543; J.L. Halaas, K.S. Gajiwala, M. Maffei, et al., Weight-reducing effects of the plasma protein encoded by the *obese* gene, *Science* 269(1995):543–546; L.A. Campfield, F.J. Smith, and Y. Guisez, Recombinant mouse OB protein: Evidence for a peripheral signal linking adiposity and central neural networks, *Science* 269(1995):546–549.

p. 182 **dozens of diets they'd tried:** S.B. Heymsfield, A.S. Greenberg, K. Fujioka, et al., Recombinant leptin for weight loss in obese and lean adults: a randomized, controlled, dose-escalation trial, *Journal of the American Medical Association* 282(1999):1568–1575; M. Chicurel, Whatever happened to leptin? *Nature* 404(2000):538–540.

p. 182 **the effects of leptin as well:** J. Marx, Cellular warriors at the battle of the bulge, *Science* 299(2003):846–849.

p. 182 **known as the arcuate nucleus . . . a loss of appetite and weight loss:** Ibid.; M.W. Schwartz, S.C. Woods, D. Porte, et al., Central nervous system control of food intake, *Nature* 404(2000):661–671; J. Flier and E. Maratos-Flier, Energy homeostasis and body weight, *Current Biology* 10(2001):R215–R217.

p. 183 **and the pounds stay on:** J.M. Friedman, Obesity in the new millennium, *Nature* 404(2000):632–634.

p. 183 **without aggressive intervention:** Friedman (2003); Friedman (2000).

p. 182 **in different environments:** Friedman (2003).

p. 183 **"depends on the premium placed on obesity":** Friedman notes that his argument is an elaboration of that advanced by biogeographer and author Jared Diamond. Interested readers are encouraged to consult Diamond's 1997 book *Guns, Germs and Steel: The Fates of Human Societies* (New York: Norton) or J. Diamond, The double puzzle of diabetes, *Nature* 423(2003):599–602 for more information.

p. 183 **"In modern times, obesity and leptin resistance":** Friedman (2003).

5
"THE SCENARIO-BUFFERED BUILDING"

p. 185 **"Buildings keep being pushed around":** S. Brand, *How Buildings Learn: What Happens After They're Built* (New York: Viking, 1994), p. 5.

p. 186 **"They're designed not to adapt":** Ibid., p. 2.

p. 186 **will have to do to accommodate those developments:** Ibid., pp. 178–189.

p. 186 **"A good strategy ensures":** Ibid., p. 178. Notable examples of buildings that implement a user-friendly strategy include a number of prominent scientific research facilities. For example, Brand describes the Lewis Thomas Molecular Biology Laboratory at Princeton University, built in 1986, as "A . . . research building that works very well, thanks to intelligent programming. . . . The generous width of the corridors, designed for conviviality, turned out to be essential for materials-handling on flatbeds and glassware carts (that was a surprise not designed for), and the labs designed to share technical equipment—for conviviality as well as economy—were able to absorb many more users than expected when the building population grew from the anticipated 100 people to 300." Tony Hunter, a signaling researcher at the University of California San Francisco, cites the Salk Institute in La Jolla as "a good example of a research building where all the space is totally flexible because the laboratory floors have no internal structural walls (the services are in a separate floor above each working floor) and can therefore be configured in any way that is needed—and has been over the 40 years since the building was opened."

p. 188 **a new type of cell, the neuron:** Neurons, in fact, are probably one of the oldest types of specialized cells in multicellular animals, present in all metazoans except the sponges (Phylum Porifera).

p. 190 **And in the Spanish city . . . our understanding of neuronal communication:** S. Ramon y Cajal, *Recollections of My Life*, translated by E. Horne Craigie with J. Cano (Cambridge, MA: MIT Press, 1996), p. 251.

p. 190 **in the laboratory of his histology professor:** Ibid., pp. 249–252.

p. 190 **such a scurrilous profession:** S. Finger, *Minds Behind the Brain: A History of the Pioneers and Their Discoveries*, (New York: Oxford University Press, 2000), p. 207.

p. 191 **Gleaning what he could from journals . . . illustrated a histology textbook:** Cajal, *Recollections*, pp. 275–303.

p. 191 **"In my systematic explorations":** Ibid., p. 304.

p. 191 **"To know the brain":** Ibid., p. 305.

p. 191 **"Nobody could answer this simple question":** Ibid.

p. 191 **what it was that inspired these individuals:** Finger, *Minds*, pp. 301–310.

p. 191 **"The truly great scientists":** Ibid., p. 308.

p. 192 **"He treated the microscopic scene":** W.M. Cowan and E.R. Kandel, A brief history of synapses and synaptic transmission, in W.M. Cowan, T.C. Südhof, and C.F. Stevens, *Synapses* (Baltimore: The Johns Hopkins University Press, 2001), pp. 1–87.

p. 192 **the German biologists Jacob Schlieden and Theodor Schwann:** Ibid.; Finger, *Minds*, p. 201; C. Sotelo, Viewing the brain through the master hand of Ramón y Cajal, *Nature Reviews Neuroscience* 4(2003):71–77.

p. 193 **a new technique he had learned in Paris:** Cowan and Kandel (2001); Sotelo (2003); Cajal, *Recollections*, pp. 307–309; Finger, *Minds*, p. 208.

p. 193 **more detailed images of neurons:** Finger, *Minds*, pp. 203–205.

p. 193 **somewhere between 1 and 3 percent:** Ibid.; Cowan and Kandel (2003).

p 193 **the "tool of revelation":** Cajal, *Recollections*, p. 308.

p. 194 **"Two methods come to mind":** Ibid., p. 324.

p. 194 **"the nerve cells, which are still relatively small":** Ibid.

p. 194 **like "ivy or lianas to the trunks of trees":** Ibid., p. 332.

p. 194 **two historic conclusions:** Cowan and Kandel (2001); Finger, *Minds*, pp. 210–212; R.R. Llinás, The contribution of Santiago Ramón y Cajal to functional neuroscience, *Nature Reviews Neuroscience* 4(2003):77–80.

p. 195 **stay at his house afterward:** Cowan and Kandel (2003); Cajal, *Recollections*, pp. 417–419.

p. 195 **"it is convenient to have a term":** Cowan and Kandel (2003).

p. 196 **the gut or the adrenal gland:** P. De Camilli, V. Hauke, K. Takei, et al., The structure of synapses, in W.M. Cowan, T.C. Südhof, and C.F. Stevens, *Synapses* (Baltimore: The Johns Hopkins University Press, 2001), pp. 89–133.

p. 196 **refilling them locally:** P. de Camilli, V.I. Slepnev, O. Shupliakov, et al., Synaptic vesicle endocytosis, in W.M. Cowan, T.C. Südhof, and C.F. Stevens, *Synapses* (Baltimore: The Johns Hopkins University Press, 2001), pp. 217–274.

p. 197 **at a moment's notice:** De Camilli, et al. (2001); K.M. Harris and P. Sultan, Variation in the number, location, and size of synaptic vesicles provides an anatomical basis for the nonuniform probability of release at CA1 hippocampal synapses, *Neuropharmacology* 34(1995):1387–1395.

p. 198 **pawing in the starting gate:** De Camilli, et al. (2001); F.E. Bloom and G.K. Aghajanian, Cytochemistry of synapses: Selective staining for electron microscopy, *Science* 154(1966):1575–1577; A. Peters, S.L. Palay, and H. deF. Webster, *The Fine Structure of the Nervous System : The Neurons and Supporting Cells* (Philadelphia: Saunders, 1976), pp. 132–135.

p. 198 **concentrated at the active zone:** De Camilli, et al. (2001); N. Hirokawa, K. Sobue, K. Kanda, et al., The cytoskeletal architecture of the presynaptic terminal and molecular structure of synapsin 1, *Journal of Cell Biology* 108 (1989): 111–126.

p. 198 **to channel calcium ions:** T.C. Südhof and R.H. Scheller, Mechanism and regulation of neurotransmitter release, in W.M. Cowan, T.C. Südhof, and C.F. Stevens, *Synapses* (Baltimore: The Johns Hopkins University Press, 2001), pp. 177–215.

p. 198 **apparatus of the presynaptic neuron:** De Camilli, et al. (2001); Peters, et al., *The Fine Structure of the Nervous System*, pp. 136–142.

p. 198 **by the type known as "AMPA":** V.N. Kharazia and R.J. Weinberg, Tangential synaptic distribution of NMDA and AMPA receptors in rat neocortex, *Neuroscience Letters* 238(1997):41–44.

p. 198 **hooks and loops of Velcro:** T.C. Südhof, The synaptic cleft and synaptic cell adhesion, in W.M. Cowan, T.C. Südhof, and C.F. Stevens, *Synapses* (Baltimore: The Johns Hopkins University Press, 2001), pp. 275–313.

p. 199 **identifying an appropriate partner:** A.M. Fannon and D.R. Colman, A model for central synaptic junctional complex formation based on the differential adhesive specificities of the cadherins, *Neuron* 17(1996):423–434.

p. 199 **"parasynaptic" cadherins:** Südhof (2001).

p. 199 **was not actually awarded:** V. Hamburger, S. Ramón y Cajal, R.G. Harrison, and the beginnings of neuroembryology, *Perspectives in Biology and Medicine* 23(1980):600–616, available online from the *Developmental Biology* website at http://www.devbio.com/article.php?ch=13&id=145; J.A. Schiff, "An Unsung Hero of Medical Research," *Yale Alumni Magazine*, February 2000, available online at http://www.yalealumnimagazine.com. "America's most famous unknown scientist," according to an article in *Fortune* (notes Schiff) and certainly one of the unluckiest, Harrison was nominated for the Nobel a second time, in 1935. This time he was passed over in favor of fellow developmental biologist Hans Spemann, on the grounds that "opinions diverged and in view of the rather limited value of the method . . . an award [to Harrison] could not be made." Harrison's "limited method"—tissue culture—has become, of course, one of the mainstays of the modern cell biology laboratory.

p. 199 **also intrigued the master:** Hamburger (1980); Cajal, *Recollections*, pp. 365–370.

p. 200 **a ring of wax on a microscope slide:** G.W. Gray, The organizer, *Scientific American* 197(1957):79–88.

p. 200 **"appeared as a concentration of protoplasm":** Cajal, *Recollections*, p. 369.

p. 201 **possessing "an exquisite chemical sensitivity":** Ibid.

p. 202 **set about building a synapse:** J. R. Sanes and T. M. Jessell, The guidance of axons to their targets, in E.R. Kandel, J.H. Schwartz, and T.M. Jessell, ed., *Principles of Neural Science*, 4th ed. (New York: McGraw-Hill, 2000), pp. 1063–1086.

p. 202 **opposite side of the spinal cord:** M. Tessier-Levigne and C.S.Goodman, The molecular biology of axon guidance, *Science* 274(1996):1123–1132; S.F. Gilbert, *Developmental Biology*, 7th ed. (Sunderland, MA: Sinauer Associates, 2003), p. 447.

p. 204 **"netrins" which direct the growth cone:** Sanes and Jessell (2000); Tessier-Levigne and Goodman (1996); T. Serafini, T.E. Kennedy, M.J. Galko, et al., The netrins define a family of axon-outgrowth-promoting proteins homologous to *C. elegans* UNC-6, *Cell* 78(1994): 49–360; T.E. Kennedy, T. Serafini, J.R. de la Torre, et al., Netrins are diffusible chemotropic factors for commissural axons in the embryonic spinal cord, *Cell* 78(1994):425–435; U. Drescher, Netrins find their receptor, *Nature* 384(1996):416–417; T.E. Kennedy, Cellular mechanisms of netrin function; Long-range and short-range function, *Biochemistry and Cell Biology* 78(2000):569–575; X. Li, M. Meriane, I. Triki, et al., The adaptor protein Nck-1 couples the netrin-1 receptor DCC (deleted in colorectal cancer) to the activation of the small GTPase Rac1 through an atypical mechanism, *Journal of Biological Chemistry* 277(2002):37788–37797. The name "netrin," incidentally, comes from a Sanskrit word *netr*, meaning "one who guides" (H. Lodish, A. Berk, S.L. Zipursky, et al., *Molecular Cell Biology*, 4th. ed. [New York: W.H. Freeman, 2000], p. 1036).

p. 204 **toward the center of the floor plate:** Gilbert, *Developmental Biology*, p. 447; Serafini, et al. (1994); Drescher (1996); S.A. Colamarino and M. Tessier-Levigne, The role of the floor plate in axon guidance, *Annual Review of Neuroscience* 18(1995):497–529.

p. 204 **the bone morphogenetic proteins:** F. Charron, E. Stein, J. Jeong, et al., The morphogen Sonic hedgehog is an axonal chemoattractant that collaborates with netrin-1 midline axon guidance, *Cell* 113(2003):11–23; A. Augsberger, A. Schuchardt, S. Hoskins, et al., BMPs as mediators of roof plate repulsion of commissural neurons, *Neuron* 24(1999):127–141.

p. 204 **"Shh and BMPs, which initially cooperate":** Charron, et al. (2003).

p. 204 **rolled out for them across the midline:** Tessier-Levigne and Goodman (1996).

p. 205 **matrix that coats their surfaces:** J. Cohen, J.F. Burne, C. McKinlay, et al., The role of laminin and the laminin/fibronectin receptor complex in the outgrowth of retinal ganglion axons, *Developmental Biology* 122(1987):407–418; P. Liesi and J. Silver, Is astrocyte laminin involved in axon guidance in the mammalian CNS? *Developmental Biology* 130(1988):774–785.

p. 205 **to intracellular signal transduction relays:** Sanes and Jessell (2000a); R.O. Hynes, Integrins: Bidirectional allosteric signaling machines, *Cell* 110(2002): 673–687; F.G. Giancotti and E. Ruoslahti, Integrin signaling, *Science* 285(1999): 1028–1032.

p. 205 **comparable to Gα:** Hynes (2002).

p. 205 **a short tail that spans the membrane:** Hynes (2002); Giancotti and Ruoslahti (1999).

p. 205 **that allow them to read it:** Cohen, et al. (1987); Liesi and Silver (1988).

p. 206 **ephrin A2 and A5:** Sanes and Jessell (2000a); J. Frisen, P.A. Yates, T. McLaughlin, et al., Ephrin-A5 (AL-1/RAGS) is essential for proper retinal axon guidance and topographic mapping in the mammalian visual system, *Neuron* 20(1998): 233–243; J.G. Flanagan and P. Vanderhaeghen, The ephrins and EPH receptors in neural development, *Annual Review of Neuroscience* 21(1998):309–345.

p. 206 **backwards into the signaling cell:** A.W. Boyd and M. Lackmann, Signals from Eph and Ephrin proteins: A developmental tool kit, *Science's STKE* (2001):1–6, available online at www.stke.org/cgi/content/full/OC_sigtrans;2001/112/re20.

p. 206 **retreats to the ephrin-free anterior tectum . . . the posterior tectum's ephrin insults:** Sanes and Jessell (2002a); J. Walter, S. Henke-Fahle, and F. Bonhoeffer, Avoidance of posterior tectal membranes by temporal retinal axons, *Development* 101 (1987):909–913; Frisen, et al. (1998); Flanagan and Vanderhaeghen (1998).

p. 206 **on its way to a lifetime partnership:** In the developing brain, repulsion and attraction are not mutually exclusive. On city streets, red always means stop and green always means go, but on the extracellular highways traveled by emergent axons and infant neurons, some signals can mean both. Take a trip through the trochlear nerve that controls eye movements, for example; the limbic system, central to the regulation of emotional behavior; or the cerebral cortex during their development and you're certain to hear some form of the word "semaphorin," the collective name for seven distinct classes of signaling molecules featuring a common protein module, the sema domain. To many listeners, the semaphorins are variations on "no." But to young pyramidal neurons in the cortex, one, semaphorin 3A, means "come here" as well as "go away." Located three, five, or even six cell layers below the convoluted cortical surface, these cells face a unique developmental challenge: each has not only an axon, but also a distinctive elongated dendrite which must find its way to targets in the more superficial layers. Both processes rely on a top-to-bottom gradient of semaphorin 3A to navigate. But while pyramidal axons respond in the negative, growing away from the source of the semaphorin 3A signal, the so-called apical dendrites stretch up toward it like seedlings seeking sunlight.

Semaphorin 3A means different things to different parts of a pyramidal cell because pyramidal axons and apical dendrites complete a semaphorin sentence two different ways. Dendrites, but not axons, contain an intracellular signaling molecule, guanylate cyclase. Only dendrites, therefore, can insert this word after "neuropilin-1"(the semaphorin 3A receptor)—and that little addition, apparently, is enough to transform semaphorin 3A signal from a repellant into an attractant. See F. Polleux, T. Morrow, and A. Ghosh, Semaphorin 3A is a chemoattractant for cortical apical dendrites, *Nature* 404(2000):567–573. For more information on the role of semaphorins as axon guidance cues in the central nervous system, see A. Sahay, M.E. Molliver, D.D. Ginty, et al., Semaphorin 3F is critical for the development of limbic system circuitry and is required in neurons for selective CNS axon guidance events, *Journal of Neuroscience* 23(2003):6671–6680.

p. 208 **take charge of its life:** S. Cohen-Cory, The developing synapse: Construction and modulation of synaptic structures and circuits, *Science* 298(2002):770–776; J.R. Sanes and T.M. Jessell, The formation and regeneration of synapses, in E.R. Kandel, J.H. Schwartz, and T.M. Jessell, ed., *Principles of Neural Science*, 4th ed. (New York: McGraw-Hill, 2000), pp. 1087–1114; A.M. Craig and J.W. Lichtman, Synapse formation and maturation, in W.M. Cowan, T.C. Südhof, and C.F. Stevens, *Synapses* (Baltimore: The Johns Hopkins University Press, 2001), pp. 571–612.

p. 208 **a cytoskeletal protein called rapsyn:** Sanes and Jessell (2000b); J.R. Sanes and J.W. Lichtman, Development of the vertebrate neuromuscular junction, *Annual Review of Neuroscience* 22(1999):389–442.

p. 208 **to fill out each cluster:** A.W. Sandrock, S.E. Dryer, K.M. Rosen, et al., Maintenance of acetylcholine receptor number by neuregulins, *Science* 276(1997):599–603.

p. 208 **receptor making or gathering:** Sanes and Jessell (2000b).

pp. 208–209 **reinforce the protein scaffolding:** Ibid.

p. 209 **grow up and start releasing neurotransmitter:** Ibid.; B.L. Patton, J.H. Miner, A.Y. Chiu, et al., Localization, regulation and function of laminins in the neuromuscular system of developing, adult and mutant mice, *Journal of Cell Biology* 139(1997):1507–1521.

p. 209 **growth and survival factors:** Sanes and Jessell (2000b).

p. 209 **acetylcholine receptors in muscle fibers:** Ibid.; Cohen-Cory (2002); Craig and Lichtman (2001).

p. 209 **the mystery signals collect them:** Sanes and Jessell (2000b); Craig and Lichtman (2001). The name of the protein is "gephyrin," and its similarity to rapsyn is strictly functional; it is structurally unrelated to the muscle protein.

p. 209 **time to assemble synaptic vesicles:** P. Scheiffele, J. Fan, J. Choih, et al., Neuroligin expressed in nonneuronal cells triggers presynaptic development in contacting axons, *Cell* 101(2000):657–669.

p. 209 **all but one axon withdrawn:** Craig and Lichtman (2001); J.W. Lichtman and H. Colman, Synapse elimination and indelible memory, *Neuron* 25(2000):269–278.

p. 212 **stimulates the postsynaptic cell to action:** M. H.-T. Sheng, The postsynaptic specialization, in W.M. Cowan, T.C. Südhof, and C.F. Stevens, *Synapses*

(Baltimore: The Johns Hopkins University Press, 2001), pp. 315–355; M.B. Kennedy, Signal-processing machines at the postsynaptic density, *Science* 290(2000):750–754.

p. 212 **earliest stages of synapse formation:** C. Lüscher, R.A. Nicoll, R.C. Malenka, et al., Synaptic plasticity and dynamic modulation of the postsynaptic membrane, *Nature Neuroscience* 3(2000):545–550.

p. 212 **AMPA receptors . . . calcium as well as sodium and potassium:** E. R. Kandel and S.A. Siegelbaum, Synaptic integration, in E.R. Kandel, J.H. Schwartz, and T.M. Jessell, ed., *Principles of Neural Science,* 4th ed. (New York: McGraw-Hill, 2000), pp. 207–227.

p. 213 **widespread than calcium:** M.J. Berridge, M.D. Bootman, and P. Lipp, Calcium—A life and death signal, *Nature* 395(1998):645–648; B. Alberts, A. Johnson, J. Lewis, et al., *Molecular Biology of the Cell,* 4th ed. (New York, Garland Science, 2002), pp. 861–865.

p. 213 **The most ubiquitous . . . is called calmodulin:** Alberts, et al., *Molecular Biology of the Cell,* pp. 863–865.

p. 213 **CAM kinases for short:** Ibid.

p. 213 **of only 20,000 neurons:** E.R. Kandel, Cellular mechanisms of learning and the biological basis of individuality, in E.R. Kandel, J.H. Schwartz, and T.M. Jessell, ed., *Principles of Neural Science,* 4th ed. (New York: McGraw-Hill, 2000), pp. 1247–1279; E.R. Kandel, The molecular biology of memory storage: A dialogue between genes and synapses, *Science* 294 (2001):1030–1038.

p. 213 **for days or even weeks:** Ibid.

p. 214 **synapses on a sensory neuron . . . and regulates its activity:** Kandel (2000).

p. 214 **the activation of protein kinase A:** Ibid.; Kandel (2001); J.H. Byrne and E.R. Kandel, Presynaptic facilitation revisited: State and time dependence, *Journal of Neuroscience* 16(1996):425–435.

p. 215 **it complains to the MAP kinase:** Kandel (2000); Kandel (2001); D. Michael, K.C. Martin, R. Seger, et al., Repeated pulses of serotonin required for long-term facilitation activate mitogen-activated protein kinase in sensory neurons of *Aplysia, Proceedings of the National Academy of Sciences* 95(1998):1864–1869.

p. 215 **the addition of new synapses:** C.H. Bailey and E.R. Kandel, Structural changes accompanying memory storage, *Annual Review of Physiology* 55(1993):397–426.

p. 215 **"long term potentiation," or LTP:** T.V.P. Bliss and T. Lomø, Long-lasting potentiation of synaptic transmission in the dentate gyrus of the anesthetized rabbit following stimulation of the perforant path, *Journal of Physiology* (London) 232(1973):331–356.

p. 216 **Like sensitization in *Aplysia* . . . an effect that lasts two to three hours:** Lüscher, et al. (2000); Kandel (2000); Kandel (2001); C. Lüscher, et al., Role of AMPA receptor cycling in synaptic transmission and plasticity, *Neuron* 24(1999): 649–658; S.H. Shi, et al., Rapid spine delivery and redistribution of AMPA receptors after synaptic NMDA receptor activation, *Science* 284(1999):1811–1816; R.C. Malenka and S.A. Siegelbaum, Synaptic plasticity: Diverse targets and mechanisms for regulating synaptic efficacy, in W.M. Cowan, T.C. Südhof, and C.F. Stevens, *Synapses* (Baltimore: The Johns Hopkins University Press, 2001), pp. 393–453.

So-called silent synapses—excitatory synapses that lack functional AMPA receptors at the surface but have such receptors sequestered within the postsynaptic neuron—can be converted to functional synapses by LTP induction, presumably because the induction message includes a directive to move the internal AMPA receptors to the surface of the postsynaptic membrane.

p. 216 **Once again . . . that will last at least 24 hours:** Lüscher, et al. (2000); Kandel (2000); Kandel (2001); Malenka and Siegelbaum (2001); R.C. Malenka and R.A. Nicoll, Long-term potentiation—A decade of progress? *Science* 285(1999):1870–1874; M. Sheng and M.J. Kim, Postsynaptic signaling and plasticity mechanisms, *Science* 298(2002):776–780.

p. 217 **carry out local protein synthesis:** K.C. Martin, M. Barad, and E.R. Kandel, Local protein synthesis and its role in synapse-specific plasticity, *Current Opinion in Neurobiology* 10(2000):587–592; J.P. Pierce, K, van Leyen, and J.B. McCarthy, Translocation machinery for synthesis of integral membrane and secretory proteins in dendritic spines, *Nature Neuroscience* 3(2000):311–313.

p. 217 **will now be allowed to use them:** Kandel (2001); Sheng and Kim (2002); Martin, et al. (2000); A. Casadio, K.C. Martin, M. Giustetto, et al., A transient, neuron-wide form of CREB-mediated long-term facilitation can be stabilized at specific synapses by local protein synthesis, *Cell* 99(1999):221–237; K.C. Martin, A. Casadio, H. Zhu, et al., Synapse-specific, long-term facilitation of *Aplysia* sensory to motor synapses: A function for local protein synthesis in memory storage, *Cell* 91(1997):927–938.

p. 218 **"the anatomical equivalent of city hall":** D.L. Niehoff, *The Biology of Violence: How Understanding the Brain, Behavior, and Environment Can Break the Vicious Circle of Aggression* (New York, The Free Press, 1999), p. 96.

p. 218 **older memories (courtesy of the hippocampus) intersect:** Ibid.; E. Halgren, Emotional physiology of the amygdala within the context of human cognition, in A.P. Aggleton, ed., *The Amygdala: Neurobiological Aspects of Emotion, Memory, and Mental Dysfunction* (New York, Wiley-Liss, 1992), pp. 191–228.

p. 219 **its heart pounding and its blood pressure soaring:** J.E. LeDoux, Emotion, memory, and the brain, *Scientific American* 270(1994):32–39; J.E. LeDoux, Information flow from sensation to emotion: Plasticity in the neural computation of stimulus value, in M. Gabriel and J. Moore, *Neuroscience: Foundation of Adaptive Networks* (Cambridge, MA: MIT Press, 1990), pp. 3–51; J.E. LeDoux, In search of an emotional system in the brain: Leaping from fear to emotion and consciousness, paper presented at the McDonnell-Pew Conference on Cognitive Neuroscience, Lake Tahoe, CA, 1993.

p. 219 **can reawaken it:** Le Doux (1990).

p. 219 **induces LTP in the hippocampus:** R.C. Malenka and R.A. Nicoll, Never fear, LTP is hear, *Nature* 390 (1997): 552–553; M.T. Rogan, U.V. Stäubli, and J.E. LeDoux, Fear conditioning induces associative long-term potentiation in the amygdala, *Nature* 390(1997):604–607; M.G. McKernan and P. Shinnick-Gallagher, Fear conditioning induces a lasting potentiation of synaptic currents in vitro, *Nature* 390(1997):607–611.

p. 219 **exposed to the tone but never shocked:** Rogan, et al. (1997).

p. 220 **"If you want a building to learn":** Brand, *How Buildings Learn*, p. 190.

p. 220 **Denise is another example:** Not her real name.

p. 220 **including alterations in gene expression:** S.E. Hyman and E.J. Nestler, Initiation and adaptation: A paradigm for understanding psychotropic drug action, *American Journal of Psychiatry* 153(1996):151–162.

p. 221 **changes in the structure and function of neurons:** E.J. Nestler, Molecular basis of addictive states, *The Neuroscientist* 1(1995):212–220; S.E. Hyman and R.C. Malenka, Addiction and the brain: The neurobiology of compulsion and its persistence, *Nature Reviews Neuroscience* 2(2001):695–703; J.Chao and E.J. Nestler, Molecular neurobiology of drug addiction, *Annual Review of Medicine* 55(2004):113–132; N.D. Volkow and T.K. Li, Drug addiction: The neurobiology of behavior gone awry, *Nature Reviews Neuroscience* 5(2004):963–970. One of the proteins altered by repeated exposure to drugs of abuse is the transcription factor CREB, confidante of cyclic AMP during the late phase of long-term potentiation. The induction of addiction, in other words, appears to involve at least some of the same molecules and mechanisms as those implicated in the storage of explicit memory.

p. 221 **particularly so-called reward pathways:** Nestler (1995); Hyman and Malenka (2001); Chao and Nestler (2004); Volkow and Li (2004); A.R. Childress, P.D. Mozley, W. McElgin, et al., limbic activation during cue-induced cocaine craving, *American Journal of Psychiatry* 156(1999):11–18; G. DiChiara and A. Imperato, Drugs of abuse preferentially stimulate dopamine release in the mesolimbic system of freely moving rats, *Proceedings of the National Academy of Sciences* 85(1988):5274–5278.

p. 222 **"Not all adaptations":** S.E. Hyman, presentation at the American Psychiatric Association Annual Meeting, May 1996.

p. 224 **synonyms for "help":** C. Nathan, Points of control in inflammation, *Nature* 420(2002):846–852.

p. 224 **and devour them alive:** C.A. Janeway and P. Travers, *Immunobiology: The Immune System in Health and Disease*, 3rd ed. (New York: Garland Publishing, 1997), pp. 9:10–9:12; C.A Janeway and R. Medzhitov, Innate immune recognition, *Annual Review of Immunology* 20(2000):197–216; R. Medzhitov and C.A. Janeway, Innate immunity, *New England Journal of Medicine* 343(2000):338–344.

p. 224 **to issue a call for reinforcements:** Nathan (2002); Janeway and Travers, *Immunobiology*, pp. 9:13, 9:18–9:19; S. Werner and R. Grose, Regulation of wound healing by growth factors and cytokines, *Physiological Reviews* 83(2003):835–870.

p. 224 **Known collectively as chemokines:** Janeway and Travers, *Immunobiology*, pp. 9:18–9:19; J.G. Cyster, Chemokines and cell migration in secondary lymphoid organs, *Science* 286(1999):2098–2102; G. Gerard and B.J. Evans, Chemokines and disease, *Nature Immunology* (2001):108–115.

p. 225 **into the infected tissue:** Janeway and Travers, *Immunobiology*, pp. 9:15–9:16; P.J. Delves and I.M. Roitt, The immune system: Part 1, *New England Journal of Medicine* 343(2000):37–49.

p. 225 **leak into the tissue in the process:** Nathan (2002).

p. 225 **a handful of features common to . . . microorganisms:** Janeway and Travers, *Immunobiology*, pp. 9:10–9:12; Janeway and Medzhitov (2002); Medzhitov and Janeway (2000).

p. 226 **RAG proteins:** D.J. Laird, A.W. De Tomaso, M.D.Cooper, et al., 50 million years of chordate evolution: Seeking the origins of adaptive immunity, *Proceedings of the National Academy of Sciences* 97(2000):6924–6926. The authors note that the precipitous evolution of RAG-mediated adaptive immunity has had such a profound influence on the evolutionary success of the vertebrate lineage some immunologists have referred to it as the "Immunological Big Bang."

p. 226 **waiting around for evolution to do it:** Laird, et al. (2000); E.E. Max, Immunoglobulins: molecular genetics, in W.E. Paul, *Fundamental Immunology*, 4th ed. (Philadelphia: Lippincott-Raven, 1999), pp. 111–182.

p. 226 **immunoglobulins and one near relative:** The designer proteins produced by white blood cells, the immunoglobulins and the constituent subunits of the so-called T lymphocyte receptor, are members of a gargantuan family of proteins known as the "immunoglobulin (Ig) supergene family," all of which contain a characteristic structural feature called the Ig fold. In addition to these superstars of the immune system, the Ig supergene family includes several well-known cell adhesion molecules, as well as the DCC receptors that bind netrins and ephrin receptors.

p. 226 **millions of bacterial and viral features:** Janeway and Travers, *Immunobiology*, p. 1:15.

p. 227 **Lymphocytes come in two flavors:** Delves and Roitt (2000a); W.E. Paul, The immune system: An introduction, in W.E. Paul, *Fundamental Immunology*, 4th ed. (Philadelphia: Lippincott-Raven, 1999), pp. 1–18.

p. 227 **but they are not immunoglobulins:** Paul (1999).

p. 227 **sent to boarding school in the thymus:** Detailed discussions of the selection, development, and activation of B and T lymphocytes can be found in any standard textbook of immunology. See, for example, chapters 5–8 in Janeway and Travers, *Immunobiology* or chapters 6, 7, 11, and 12 in Paul, *Fundamental Immunology*.

p. 227 **commit suicide as soon as possible:** E. Palmer, Negative selection—Clearing out the bad apples from the T-cell repertoire, *Nature Reviews Immunology* 3 (2003):383–391; J.T. Opferman and S.J. Korsmeyer, Apoptosis in the development and maintenance of the immune system, *Nature Immunology* 4(2003):410–415.

p. 229 **hence their name: dendritic cells:** J. Banchereau and R.M. Steinman, Dendritic cells and the control of immunity, *Nature* 392(1998):245–252.

p. 229 **educators, mentors, motivators, counselors, and dispatchers:** Banchereau and Steinman (1998); R.M. Steinman and M.C. Nussenzweig, Avoiding horror autotoxicus: The importance of dendritic cells in peripheral T cell tolerance, *Proceedings of the National Academy of Sciences* 99(2002):351–358; R.M. Steinman, Dendritic cells, in W.E. Paul, *Fundamental Immunology*, 4th ed. (Philadelphia: Lippincott-Raven, 1999), pp. 547–573.

p. 229 **the so-called major histocompatibility complex:** Janeway and Travers, *Immunobiology*, pp. 1:24–1:25, 4:2–4:3; P.J. Delves and I.M. Roitt, The immune system, Part 2, *New England Journal of Medicine* 343(2000):108–117.

p. 229 **a so-called costimulatory signal as well:** Delves and Roitt (2000b); Janeway and Travers, *Immunobiology*, pp. 7:8–7:9. Engagement of a T cell's receptor in the absence of a costimulatory signal is a dead giveaway that the cell in question is binding to a self antigen, rather than a bacterial protein. Such misfirings trigger a fail-safe reaction known as *anergy*, in which the offending cell is stripped of its ability to respond to antigen and condemned to a permanent state of suspended animation.

p. 230 **job title is "professional antigen presenting cell":** Banchereau and Steinman (1998); Steinman (1999).

p. 230 **start looking for a partner:** Ibid.

p. 230 **a supply of the essential costimulator:** Banchereau and Steinman (1998); Steinman (1999); Steinman and M.C. Nussenzweig (2002).

p. 232 **to talk to the B cells they activated:** P.A. van der Merwe and S.J. Davis, The immunological synapse—A multitasking system, *Science* 295(2002):1479–1481; W.E. Paul and R.E. Sader, Lymphocyte responses and cytokines, *Cell* 76(1994):241–251.

p. 232 **the T cell contacted its signaling partner:** C.R. Monks, B.A. Freiberg, H. Kupfer, et al., Three-dimensional segregation of supramolecular activation clusters in T cells, *Nature* 395(1998):82–86; A. Grakoui, S.K. Bromley, C. Sumen, et al., The immunological synapse: A molecular machine controlling T cell activation, *Science* 285(1999):221–227; J.P. Roberts, Dissecting the immunological synapse, *The Scientist* 17(2003), available online at www.the-scientist.com/yr2003/may/research1_030505.html.

p. 232 **like a bull's-eye:** Monks, et al. (1998); Grakoui, et al. (1999); Roberts (2003); M.L. Dustin and A.S. Shaw, Costimulation: Building an immunological synapse, *Science* 283(1999):649–651; J. Delon, The immunological synapse, *Current Biology* 10(2003):214; M.L. Dustin and D.R. Colman, Neural and immunological synaptic relations, *Science* 298(2002):785–789.

p. 232 **the active zone in a neural synapse:** Grakoui, et al. (1999); Roberts (2003); Delon (2003); Dustin and Colman (2002); C. Wülfing and M.M. Davis, A receptor/cytoskeletal movement triggered by costimulation during T cell activation, *Science* 282(1998):2266–2269.

p. 233 **before construction of the synapse is finished:** van der Merwe and Davis (2002); Roberts (2003); K-H. Lee, A.D. Holdorf, M.L. Dustin, et al., T cell receptor signaling precedes immunological synapse formation, *Science* 295(2002):1539–1542.

p. 233 **to match the level of antigen:** K-H. Lee, A.R. Dinner, C. Tu, et al., The immunological synapse balances T cell receptor signaling and degradation, *Science* 302(2003):1218–1222.

p. 234 **later stages in the T cell activation process:** Dustin and Colman (2002).

p. 234 **chatting like social equals:** D.T. Fearon and R.M. Locksley, The instructive role of innate immunity in the acquired immune response, *Science* 272(1996):50–54.

p. 234 **are likely to be pathogenic:** Banchereau and Steinman (1998).

p. 234 **to the infection site:** Nathan (2002).

p. 234 **from attack mode to healing mode:** Ibid.; T.R. Mosmann, H. Cherwinski, M.W. Bond, et al., Two types of murine helper T cell clone. I. Definition according to profiles of lymphokine activities and secreted proteins, *Journal of Immunology* 136(1986):2348–2357; H.L. Weiner and D.J. Selkoe, Inflammation and therapeutic vaccination in CNS diseases, *Nature* 420(2002):879–883.

p. 235 **has had to sacrifice infallibility:** Fearon and Locksley (1996).

p. 235 **picked up self antigens during their travels:** Banchereau and Steinman (1998); Steinman and M.C. Nussenzweig (2002).

p. 235 **limit their interactions with the immune system:** Janeway and Travers, *Immunobiology*, p. 12:32.

p. 236 **known collectively as myelin basic protein:** J.H. Schwartz and G.L. Westbrook, The cytology of neurons, in E.R. Kandel, J.H. Schwartz, and T.M. Jessell, ed., *Principles of Neural Science*, 4th ed. (New York: McGraw-Hill, 2000), pp. 67–87.

p. 236 **that body has chosen to ignore:** R. Hohlfeld and H. Wekerle, Autoimmune concepts of multiple sclerosis as a basis for selective immunotherapy: From pipe dreams to (therapeutic) pipelines, *Proceedings of the National Academy of Sciences* 101(2004):14599–14606.

p. 237 **the smallest, most lipid-loving molecules:** R. Hohlfeld, Biotechnological agents for the immunotherapy of multiple sclerosis: Principles, problems, and perspectives, *Brain* 120(1997):865–916; H. Wekerle, C. Linington, H. Lassmann, et al., Cellular immune reactivity in the CNS, *Trends in Neurosciences* 9(1986):271–277.

p. 237 **activates myelin-reactive T cells:** Hohlfeld and Wekerle (2004); Hohlfeld (1997); R.T. Johnson, The virology of demyelinating diseases, *Annals of Neurology* 36(1994):S54–S60.

p. 237 **in the brain, spinal cord, and optic nerves:** Hohlfeld and Wekerle (2004); Hohlfeld (1997).

p. 237 **described this autoimmune catastrophe in 1868:** D.A. Hafler, Multiple sclerosis, *Journal of Clinical Investigation* 113(2004):788–794.

p. 237 **between the ages of 20 and 40:** Hohlfeld and Wekerle (2004); R.D. Adams, M. Victor, and A.H. Ropper, *Principles of Neurology*, 6th ed. (New York: McGill-Hill, 1997), p. 906.

p. 237 **in the Th1 direction:** Weiner and Selkoe (2002); R.R. Voskuhl, R. Martin, C. Bergman, et al., T helper (Th1) functional phenotype of human myelin basic protein-specific T lymphocytes, *Autoimmunity* 15(1993):137–143.

pp. 237–238 **precede the attack by several weeks:** P. Rieckmann, M. Albrecht, B. Kitze, et al., Tumor necrosis factor-alpha messenger RNA expression in patients with relapsing-remitting multiple sclerosis is associated with disease activity, *Annals of Neurology* 37(1995):82–88.

p. 238 **an MS patient is likely to be:** M.K. Sharief and R. Hentges, Association between tumor necrosis factor-alpha and disease progression in patients with multiple sclerosis, *New England Journal of Medicine* 325(1991):467–472.

p. 238 **the rate of disease progression:** Ibid.

p. 238 **endothelial cells don adhesive molecules:** H.P. Hartung, K. Reiners, J.J. Archelos, et al., Circulating adhesion molecules and tumor necrosis factor receptor in multiple sclerosis: Correlation with magnetic resonance imaging, *Annals of Neurology* 38(1995):186–193.

p. 238 **were intent on attacking:** D.H. Miller, O.A. Khan, W.A. Sheremata, et al., A controlled trial of natalizumab for relapsing multiple sclerosis, *New England Journal of Medicine* 348(2003):15–23.

p. 239 **a 2-year clinical trial in more than 900 MS patients:** "FDA grants accelerated approval of Tysabri® formerly Antegren® for the treatment of multiple sclerosis," press release available online, along with additional information about Tysabri®, on the Biogen Website: www.biogen.com.

p. 239 **sales were suspended . . . "But I knew it was going to happen":** "Biogen

Idec and Elan Announce Voluntary Suspension of Tysabri®", press release available online at www.elan.com/News/full.asp?ID=679361; A.W. Pollack, "Sales Halted in Biotech Drug Because of Link to a Death," *New York Times*, March 3, 2005, Business Day section.

p. 239 **"survival is impossible":** K.J. Tracey, The inflammatory reflex, *Nature* 420(2002):853–859.

p. 239 **"'Change is suffering' was the insight":** Brand, *How Buildings Learn*, p.167.

6
THE VIRTUAL CELL

p. 242 **"the big question of how they all work together":** A. Abbott, Alliance of US labs plans to build map of cell signaling pathways, *Nature* 402(1999):219–220.

p. 242 **"Sequencing the genome is enabling us":** K. Devine, Cell signaling alliance gets underway, *The Scientist* 14(2000):1,12.

p. 242 **catalog and publish as an electronic database:** Responsibility for coordinating peer review and publication of the Molecule Pages has been assumed by the editors of the journal *Nature*. In collaboration with the AfCS, *Nature* also maintains an online resource, the Signaling Gateway (www.signaling-gateway.org), featuring news updates, review articles, and links to the scientific literature. More information about the AfCS, including regular newsletters highlighting recent progress and news from the Alliance's annual meeting, can be accessed via the Signaling Gateway as well.

p. 243 **The annual budget . . . totals around $10 million:** Abbott (1999).

p. 245 **"signaling pathways interact with one another":** U.S. Bhalla and R. Iyengar, Emergent properties of networks of biological signaling pathways, *Science* 283(1999):381–387.

p. 246 **"Complexities can only be understood":** J. Maddox, *What Remains to Be Discovered: Mapping the Secrets of the Universe, the Origins of Life, and the Future of the Human Race* (New York: The Free Press, 1998), p. 184.

p. 246 **scientists realize three benefits:** E. Werner, In silico cell signaling underground, Science's STKE (2003), online at www.stke.org/cgi/content/full/sigtrans;2003/170/pe8.

p. 246 **"Models . . . can be used to plan":** Ibid.

p. 246 **"force a new perspective on the subject matter":** Ibid.

p. 246 **to appreciate the value of model building:** Maddox, *What Remains to Be Discovered*, pp. 184–189.

p. 247 **"Each word carries the whisper,":** C. Alexander, S. Ishikawa, M. Silverstein with M. Jacobson, I. Fiksdahl-King, and S. Angel, *A Pattern Language* (New York: Oxford University Press, 1977), p. xliii.

p. 248 **"a modern-day plague" . . . "a legion of recondite diseases" . . . "a terrifying alien entity":** D.R. Green and G.I. Evan, A matter of life and death, *Cancer Cell* 1(2002):19–30.

p. 248 **Herceptin, is a monoclonal antibody:** Y. Yarden, J. Baselga, and D.

Miles, Molecular approaches to breast cancer treatment, *Seminars in Oncology* 31(2004):6–13; R.S. Finn and D.J. Slamon, Monoclonal antibody therapy for breast cancer: Herceptin, *Cancer Chemotherapy and Biological Response Modifiers* 21(2003):223–233; N. Wade, "Scientists View New Wave of Cancer Drugs," *New York Times*, May 29, 2001, Science Times section.

p. 248 **extra copies of the *her2/neu* gene:** R. Roskosi, Jr., The ErbB/HER receptor protein-tyrosine kinases and cancer, *Biochemical and Biophysical Research Communications* 319(2004):1–11; J.S. Ross, J.A. Fletcher, G.P. Linette, et al., The her-2/neu gene and protein in breast cancer 2003: Biomarker and target of therapy, *Oncologist* 8(2003):307–325.

p. 248 **Gleevec silences an aberrant kinase:** B.J. Druker, Perspectives on the development of a molecularly targeted agent, *Cancer Cell* 1(2002):31–36; B.J. Druker, M. Talpaz, D. Resta, et al., Efficacy and safety of a specific inhibitor of the Bcr-Abl tyrosine kinase in chronic myeloid leukemia, *New England Journal of Medicine* 344(2001):1031–1037; B.J. Druker, C.L. Sawyers, H. Kantarjian, et al., Activity of a specific inhibitor of the Bcr-Abl tyrosine kinase in the blast crisis of chronic myeloid leukemia and acute lymphoblastic leukemia with the Philadelphia chromosome, *New England Journal of Medicine* 344(2001):1038–1042. The disrupted chromosome was discovered in 1960 by scientists at the Wistar Institute in Philadelphia and is called the "Philadelphia chromosome" as a consequence (see P.C. Nowell and D.A. Hungerford, A minute chromosome in human chronic granulocytic leukemia, *Science* 132[1960]:1497–1501). Bcr-Abl, the tyrosine kinase encoded by the fused chromosome segments is thought to stimulate proliferation and/or survival of hematopoietic progenitor cells (see M.W. Deininger, J.M. Goldman, and J.V. Melo, The molecular biology of chronic myeloid leukemia, *Blood* 96[2000]:3343–3356).

p. 248 **Gleevec blocks a mutant receptor tyrosine kinase called c-kit:** B.J. Druker and N.B. Lydon, Lessons learned from the development of an abl tyrosine kinase inhibitor for chronic myelogenous leukemia, *Journal of Clinical Investigation* 105(2000):3–7; D.A. Tuveson, N.A. Willis, T. Jacks, et al., STI571 inactivation of the gastrointestinal stromal tumor c-KIT oncoprotein; Biological and clinical implications, *Oncogene* 20(2001):5054–5058.

p. 248 **it also inhibits the platelet-derived growth factor receptor:** Druker (2002); Druker and Lydon (2000). In fact, Gleevec was originally synthesized as part of a program directed at inhibiting the platelet-derived growth factor receptor, known to be activated in the brain cancer glioblastoma and present in a number of other solid tissue tumors.

pp. 248–249 **90 percent of patients with CML . . . have their cancers go into remission:** Druker (2002); L.K. Altman, "Cancer Doctors See New Era of Optimism," *New York Times*, May 22, 2001.

p. 249 **mutations in the growth factor receptor itself:** T.J. Lynch, D.W. Bell, R. Sordella, et al., Activating mutations in epidermal growth factor receptor underlying responsiveness of non-small-cell lung cancer to gefitinib, *New England Journal of Medicine* 350(2004):2129–2139; J. Marx, Why a new cancer drug works well, in some patients, *Science* 304(2004):658–659.

p. 249 **to prevent the EGF receptor from phosphorylating itself:** Lynch, et al.

(2004); T. Hollon, The current status of cancer treatment, *The Scientist* 17(2003), available online at www.the-scientist.com/yr2003/sep/feature8_030923.html.

p. 249 **"We have only two options to collapse":** Green and Evan (2002).

p. 249 **"Alternatively," . . . "we could reinstate the defective apoptosis":** Ibid.

p. 250 **drugs designed to interfere with growth factor signaling:** Ibid.

p. 250 **only 25 to 30 percent of breast cancer patients:** Ross, et al. (2003); Wade (2001).

p. 250 **An even smaller number of lung cancer patients:** Lynch, et al. (2004); Marx (2004).

p. 250 **patients with advanced disease in the acute, or "blast-crisis" phase:** M.E. Gorre, M. Mohammed, K. Ellwood, et al., Clinical resistance to STI-571 cancer therapy caused by BCR-ABL gene mutation or amplification, *Science* 293(2001):876–880; F. McCormick, New age drug meets resistance, *Nature* 412(2001):281–282.

p. 251 **"When success does occur":** Nature Reviews Drug Discovery GPCR Questionnaire Participants, The state of GPCR research in 2004, *Nature Reviews Drug Discovery* 3(2004):577–626.

p. 251 **"We are still at the stage":** Ibid.

p. 258 **"wholeness," a quality he attributes to entities he calls "centers":** C. Alexander, *The Nature of Order: An Essay on the Art of Building and the Nature of the Universe. Book One: The Phenomenon of Life* (Berkeley, CA: The Center for Environmental Structure, 2002), pp. 79–108.

p. 258 **"a physical set that occupies a certain volume in space":** Ibid., p. 84.

p. 258 **"I notice the sunny part of the garden":** Ibid., p. 95.

p. 258 **has identified 15 fundamental structural properties:** Ibid., pp. 144–242.

p. 259 **"deep interlock and ambiguity":** Ibid., pp. 195–199.

p. 259 **"not separateness":** Ibid., pp. 230–235.

p. 259 **"Nature too is understandable in terms of wholeness":** Ibid., p. 244.

p. 259 **"the fifteen properties appear as geometric features":** Ibid., p. 246.

p. 259 **Deep interlock and ambiguity . . . appear in the pattern:** Ibid., pp. 270–271.

p. 259 **not-separateness is an integral feature of any ecosystem:** Ibid., pp. 288–289.

p. 259 **"make buildings by stringing together patterns":** Alexander, et al., *A Pattern Language*, p. xli.

p. 259 **"very dense; it has many meanings":** Ibid.

p. 260 **"buildings which are poems":** Ibid., p. xliv.

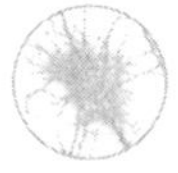

ACKNOWLEDGMENTS

In the words of Martin Rodbell, who shared the 1994 Nobel Prize in Medicine or Physiology for his work on signal transduction, this book cannot rightfully be called an individual accomplishment, but is rather the product of "a community of effort," sustained and informed by the contribution of the following scientists, who gave generously of their time and expertise to educate, guide, inspire, and correct me as I wrote and to whom I am deeply indebted: Bonnie Bassler, Philip Beachy, John Tyler Bonner, Vishva Dixit, Michael Dustin, Gerard Evan, Jeff Friedman, Alfred Gilman, E. Peter Greenberg, Stephen Hyman, C. Ronald Kahn, Eric Kandel, Mark Kirschner, Roberto Kolter, Mitch Lazar, Robert Lefkowitz, Eric Nestler, Roel Nusse, Tony Pawson, Cedric Raine, Alan Saltiel, Joshua Sanes, Solomon Snyder, Ralph Steinman, Ann Stock, Cliff Tabin, Amy Wagers, Gerry Weinmaster. In addition, I would like to thank Al Gilman, Ron Taussig, and the members of the Alliance for Cellular Signaling for affording me the opportunity to eavesdrop on their annual meeting. Tony Hunter, Asma Nusrat, and Al Gilman reviewed an early version of the manuscript in its entirety; the finished product has been much improved by their thoughtful criticism. And

I am grateful to Barbara Reynolds and Bobbi Silberg of the Diabetes Center at Saint Mary's Hospital for their help in researching the human dimension of this disorder.

Early in the evolution of this project, I realized that molecules, in contrast to words, were physical entities, meaning that sentences based on molecular interactions had a three-dimensional aspect; that they were architectural as well as linguistic constructions. Given a limited knowledge of buildings and their creation, I was fortunate to have been introduced to the work of Christopher Alexander, which proved an invaluable resource in formulating my thinking along these lines. I owe a debt as well to Scott Gilbert, without whose magnificent text on developmental biology I would have been overwhelmed by information overload.

Architecture is a visual art form, and Michael Linkinhocker did a superb job of transforming my descriptions of proteins and pathways into pictures. Jeff Robbins provided expert editorial guidance, while my agent, Regula Noetzli was also my adviser, champion, and friend throughout. Carolyn McGuiness and Haejin Chung helped with library research. Friend and fellow horse lover Marie Messerschmidt did an admirable job transcribing tapes.

The conversations between cells tell the simple story of everyday existence. Fortunately, my husband and my daughters, Jennifer and Haley, were good sports about the inclusion of examples from our everyday existence to illustrate these conversations. In addition, their collective sense of humor, disregard for domesticity, and willingness to eat macaroni and cheese on a regular basis made it possible to accomplish an otherwise impossible task. Veronica Frihart deserves a special round of applause for her skill, patience, and creativity with both human and equine students. Finally, I'd like to thank other friends and family who supported and sustained me over the course of this project: Lauren and Michael Bird, Kris Carlson, John and Caitlin Matthews, Marian Plunkett, Miriam Sexton, John and Jean Wendorf.

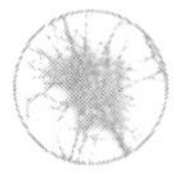

INDEX

B

C

D

E

F

G

I

J

K

L

M

N

O

P

Q

R

S

T